INTELLIGENT, AT LAST?

A COMPREHENSIVE GUIDE ON THE IMPORTANT ISSUES OF MODERN AI

NOAM ARGAMAN

CONTENTS

PREFACE

In 1589, a middle aged man decided to book an entire building in the middle of London. His mind was focused on the same thing he spent decades developing, the invention he believed would change the world entirely, and for the better. The thought of the great relief his fellow Englishmen would experience from the new machine, led him to feel gleeful and confident for his contribution to the kingdom. However, as William Lee waited to meet his monarch, Queen Elizabeth I, he knew this meeting was absolutely crucial for his future as well as the destiny of his life work, a goal he must acquire in order to conquer his dreams. Protection.

Lee's hope was to convince his queen to adopt his work and grant him a patent. His success would be solely based on the impression that he would be able to bestow on the queen. To that end, renting a centrally located building in London was a no-brainer. The goal of making a name and fortune for himself persisted above all else. Lee emptied the building from anything that was not needed, and replaced it with his grand invention—the sewing machine.

As the queen entered the building, Lee explained enthusiastically how this enormous machine works and the burden of spending years creating it, hoping to impress the queen. Acknowledging how important the queen's approval and adoption of his work was, he spared no detail. From the process itself to the benefits the kingdom could achieve from spreading his ideas across the entire country. The queen must be impressed and convinced of its importance, he thought. To Lee's consternation, the queen's attention was directed toward an unexpected issue of Lee's work. The words she uttered were far from embracing and accepting, which astonished him completely –

"Thou aimest high, Master Lee. Consider thou what the invention could do to my poor subjects. It would assuredly bring them ruin by depriving them of employment, thus making them beggars." (Acemoglu and Robinson, 2012, p. 182f)

It was quite clear that Lee's invention caused a substantial fear in the heart of the queen. The working class, the Englishmen, her subjects, would lose their jobs, if she was to approve Lee's request. By automating the process of hand-knitting, Lee pulled the rug from beneath those who were making a decent living from the knitting profession. This would bring unacceptable dangers to the queen's rule, and cause instability to her country and the future. Lee's invention was not seen as an opportunity, but as a threat.

Eventually, Lee's hopes were gone. Instead of becoming an important businessman and inventor, someone to look up to, he grew to become an outcast. His fellow Englishmen saw his life's work as a direct attack on their livelihood, since the machine could generate similar quality, substantially faster. The lack of support Lee received from his countrymen led him to flee England and travel Europe to find supporters elsewhere, which he eventually discovered in England's most notorious enemy—France. Under the patronage of King Henry IV, the stocking frame knitting machine gained popularity and contributed to the growth of the French hosiery industry.

In any case, this story may sound too familiar to you; it might be because it won't be much of a challenge to come up with several, similar stories, just like the misfortune of Master Lee. Sure, the names might be different, and some other minor details may be changed as well, but the outcome is consistent with all. A new and exciting technology threatens people's livelihood, which in turn leads to pushback, sometimes more successfully than others. As we know today, William Lee's sewing machine eventually took over those jobs, and in turn, most of the technologies we take for granted today took some other people's jobs in the past. It surely seemed unavoidable.

Could AI be our sewing machine?

In the 1900s, Albert Einstein—you might have heard of him—revolutionized our understanding of the universe with his theory of relativity. In 1905, the wonder year, he published the famous equation $E=mc^2$, which showed the relationship between energy (E) and mass (m), and how they are interchangeable.

As the world entered the 1930s, the scientific community became aware of a process known as nuclear fusion, which involves the splitting of an atomic nucleus and the release of an enormous amount of energy. In 1938, two German scientists, Otto Hahn and Fritz Strassmann, discovered the phenomenon of nuclear fusion, although they did not fully understand its implications.

Einstein, being aware of the potential power of such a discovery, wrote a letter to President Franklin D. Roosevelt in 1939, warning him about the possibility of Germany harnessing this energy for military purposes. This letter prompted the formation of the Manhattan Project, a top-secret research program tasked with developing an atomic weapon.

Led by physicist J. Robert Oppenheimer, the Manhattan Project brought together some of the world's brightest scientific minds in the United States. These scientists, including Enrico Fermi, Richard Feynman, and Niels Bohr, worked tirelessly to unlock the secrets of nuclear fission and develop a practical means of creating an atomic bomb.

Their research led to the successful construction of the world's first controlled nuclear chain reaction in December 1942, at the University of Chicago. This breakthrough demonstrated that a self-sustaining nuclear reaction was possible, paving the way for further advancements.

Meanwhile, World War II was raging on, and the Allied forces were engaged in a fierce conflict with the Axis powers. The Manhattan Project progressed rapidly, with several research facilities established across the United States, including Los Alamos, New Mexico, which became the primary site for bomb development.

By July 1945, the scientists had successfully built a working atomic bomb, and the decision was made to use it against Japan, who had not yet surrendered despite suffering significant losses. On August 6, 1945, the United States dropped the first atomic bomb, code-named "Little Boy," on the city of Hiroshima. The devastation was immense, with the blast instantly killing tens of thousands of people and causing widespread destruction.

Japan still did not surrender, and on August 9, 1945, a second atomic bomb, "Fat Man," was dropped on Nagasaki. The impact was similarly catastrophic, leading to the eventual surrender of Japan on August 15, 1945, effectively ending World War II.

Could AI be our nuclear bomb?

Answering these two questions with confidence, would indicate that the person talking might be a fool. Simply because no one really knows. We can derive some similarities in order to figure out what is common between AI and other technologies which had significant impact on the world, and what sets AI apart from the rest.

First, let's understand this.

Observing any piece of technology you could possibly name—computers, healthcare, manufacturing or space travel—every single one of them have various levels of impact on our life, based on similar levels of our understanding of a given field. When it comes to our ability to build, to create or innovate in any domain, the requirement is that we gain a certain understanding regarding the mechanics of it—how shit works, generally speaking. Imagine trying to build the first automobile, a hundred and fifty years ago without any knowledge of motors whatsoever. The smallest design for the most insignificant detail in this system requires understanding, through theory, practice or both.

In some cases, practice through trial and error is the best way to go. Consider the Wright brothers, who designed the first airplane. They did not base most of their decisions on theory or physical understanding alone—they demanded proof of concept in the field. By fixing and failing over and over again, they managed to accumulate wisdom and profound understanding of aerodynamics, second to no other human alive back at the beginning of the previous century. Of course, nowadays, engineers and physics have broadened and enriched this field with different models, capable of explaining and predicting the complex behavior of flying objects. This, like any other invention, needs a first step to revolutionize the world—understating.

To have a firm grasp on what it means to understand, it isn't necessary for us to delve into philosophy books on epistemology for years on end, at least not for the purpose of comparing AI technology to the rest. Maintaining our goal at the moment, a simple illustration would suffice.

Nicolas Copernicus was born in 1473, realizing, in 1504, that the Sun is the center around which the earth circles and not the other way around. Furthermore, Galileo, who was born in 1564, offered accurate approaches to predict the way objects move in our world while building the first telescope. Despite the achievements of these two fine gentlemen, the first scientist is considered to be Sir Isaac Newton, born in 1643, many decades after the death of the two of them. This statements' aim is not to diminish Copernicus and Galileo's contribution, but to separate their level of understanding to the one constructed by Newton—generating rejectable predictions. Ever since Newton laid out the first scientific theory, every single one of what we call scientific theory was upheld to the same standard. In order to be considered at the same level of understanding as the bar Newton created, a theory needs to provide the tools required to reject the very same theory.

Even though the influence of the scientific methodology and its effect on our everyday lives is known to all, this does not mean that the only way to create an influential and innovative discovery was built upon these concepts. After all, we were able to manufacture ships, cross oceans, erect big walls to protect us from enemies and steel cannons to destroy them, well before Newton came along. These achievements required understanding as well.

The common concept of the different levels of understanding mentioned above is the idea of modeling a complex system, condensing it to a finer, simpler and more usable representation.

Consider Newton's laws of movement, for example. By connecting between the overall forces applied on a body with its acceleration or mass, Newton is stating two separate claims, representing his theory.

First, only three parameters are needed to describe a movement of any object in the universe—the force, the mass and the acceleration. Any other variable is irrelevant. The color of the object, the material, its political beliefs and its fashion choices simply do not carry any effect regarding the manner in which the body will move.

Second, Newton claims that the way these three parameters are connected, is linear; which means that if you take the mass of the body and multiply that by the acceleration, you will surely get the force being applied on it. This might seem like an obvious observation, but the derivatives we could gather from such a claim could not be overstated. For example, we know that if we drop two objects, with the same mass from the same height, the time it will take for the two objects to reach the ground will be exactly the same. One could easily infer this simple observation from the first claim—if there are only three variables needed to describe the movement of a body, and two of these variables are equal for the two objects we examine, then it must be that the third one would be equal as well.

However, the second claim Newton made would lead to a much less intuitive discovery—the connection between these variables represents a different type of natural behavior. If we take two objects, with different mass this time, and drop them from the same height at the same time—they both would reach the ground at the exact same time. Galileo made the same prediction using a beautiful thought experiment, believed to be the first thought experiment to ever be portrayed. His analogy was to prove that if the mass is actually relevant to the time it would take for an object to reach the ground, if we are to halve the mass of the object, it would reach the ground twice as fast. Now, taking an object with a quarter of the mass would surely lead to four times the speed, and so equally one eighth of the mass would double it again. Taking one sixteenth of the mass would be great again because this way the time would be cut in half once more. This could be done over and over and over again, causing the listener to realize this may not be the case—objects simply do not behave in this manner at all.

It is obvious now how both Newton and Galileo are focusing on the same mechanism—the relationship between the relevant variables. Simply knowing which variables are relevant and which could be dropped with no hesitation cannot possibly enable us with such a profound understanding of nature.

Combining these two separate claims—which variables are relevant, and the way they relate to one another—might be the highest level of understanding we humans will ever reach. In fact, requiring these two criteria to be met is arguably what defines the scientific methodology. One might argue that knowing which variables matter and how they behave in relation to each other, necessarily leads to generating predictions that are rejectable.

Revolutionizing the world does not require meeting such a deep level, of course. The first one sometimes is enough, through trial and error. But failing both is unique. AI is the only invention which ignores both of them, completely.

Throughout this book, the mechanics and processes of how Machine Learning and AI systems work will be laid out in detail. We could look at this new and exciting technology from the prism of the definition of understanding which we just discussed, describing the world of AI as outsourcing the two steps of understanding to an unintelligent box, capable of rapidly performing simple computations, with an emphasis on the word rapidly. This means, compared to all other technologies we are familiar with, AI models are not understood like all other developments which have bene-fited our life greatly in the past few centuries. If one might consider extracting from past inventions in order to claim that the attacks we have seen recently against AI come and go all the time, there is nothing to worry about. AI would just be our sewing machine, and the people fighting against it are the modern version of English knitters, fighting a losing battle.

AI is definitely not our modern sewing machine. Currently, at least, the difference is in understanding. It might not be a catastro-phe to our jobs, but waving it off with the consistency argument is laughable. The fact that past inventions turned out to be beneficial to society and our economy, does not dictate the outcome of AI.

On the other hand, the ones who claim confidently that AI would certainly destroy all jobs simply because it mimics the way we think, lack the precise knowledge and detail of what exactly these algorithms do and the way they work. Simply labeling a new method as 'intelligent' may it be artificial or not, does not make it so. True, the origins of what we call Neural Networks were moti-vated by our knowledge of the way biological brains work and learn. However, current methodologies are so profoundly detached from the way we as humans learn, that simply connecting the lines is oversimplifying the analogy. No engineer at Google Brain or OpenAI embarks on solving a problem by dissecting a frog's brain, finding some piece of meat, yelling Eureka! It does not work that way, as simple as that. The Transformer, arguably the most import-ant architecture today, used by several famous models, is such a complex structure that no one will ever claim the design was cop-ied from or motivated by our own brains.

How AI will shape lives is yet to be determined. Basing our answers on the past or on superficial analogies would not lead to the right answer with this one, unfortunately.

After the US dropped the bomb, the entire world realized the threat embedded in this type of technology. The response was regulation—establishing the UN as well as the Atomic Energy

Commission (AEC), the NPT which entered into force in 1970, aimed to prevent the spread of nuclear weapons while promoting the peaceful use of nuclear energy, introducing the Comprehensive Nuclear-Test-Ban Treaty (CTBT) which was adopted by the UN General Assembly in 1996, the CTBT bans all nuclear explosions, including those for weapons testing. The list goes on and on.

There are many who see AI as a threat just as the atomic bomb was, and therefore, we should react to it using a similar approach. Those who champion it, led by the most successful entrepreneur of our time—Elon Musk, do not take into account two absolutely crucial differences between the threat humanity faces from AI versus nuclear bombs.

First, to build the most advanced AI system in the world right now, you will need to acquire resources—computational power, and data scientists. Both of which could be easily gathered spending a few billion dollars. The number of organizations worldwide—public and private companies, governmental institutions etc.—who can have such resources is in the thousands. Putting restrictions on all of these different players is a completely different task compared with preventing several countries that might have the knowledge and capability to construct a nuclear threat, even today.

Second, aside from the financial difference, finding out which countries aim to reach nuclear capabilities is a relatively simple task. It's just too difficult to hide. Aside from acquiring the materials needed, such as Uranium, a country exerting its best efforts to develop nuclear weapons needs to build a complex infrastructure for the development itself, for testing experiments and finally delivery platforms—missiles from the ground or from the air. All of these are not only incredibly expensive, but also fairly easy for any country with a decent intelligence force to detect. In contrast, building sophisticated AI systems is far from that—it could not be easier to keep it a secret. With the current digital revolution which transforms all of our lives to be handled by computers, acquiring massive computational power is simply not suspicious at all—

everybody is doing it for various reasons, not necessarily malicious. There is very little to no infrastructure needed for such systems besides some offices for programmers and researchers. Testing is conducted mainly in enclosed digital environments, making it impossible to discover prior to significant damage being done in an obvious manner.

AI creates immense threats and challenges to humanity, on various fronts. This technology is profoundly different from any other invention in our history. Dealing with it would require new ideas.

In this book, you will find a comprehensive and detailed analysis on the vast majority of the important concepts, tools and methods of AI, from the simplest ones designed a few decades ago to the most advanced platform you have probably heard of by now. My hope is, by providing the general public with this knowledge, more and more people could be part of the crucial conversation about the dangers of AI, which might help us solve it.

INTRODUCTION

Have you ever found yourself bewildered at just how accurately your phone responds when activated using voice commands? Or wondered how Netflix is always seamlessly able to recommend content tailored precisely toward your interests?

These seemingly magical abilities stem from transformative technologies such as machine learning, deep learning, and artificial intelligence (AI). Simply put, these cutting edge tools enable machines to learn and make decisions on their own without relying on explicit programming.

Machine Learning (ML), for example, empowers the creation of models designed to recognize various objects. One such example might involve distinguishing between images of cats and dogs. This is accomplished by inputting thousands of labeled images into a computer; from there it is free to discern key features that differentiate cats from dogs while predicting classifications for new images it has not previously encountered. Imagine presenting an unfamiliar cat image to our model; it can quickly recognize its distinguishing factors based on pointy ears, whiskers, and furry body texture—classifying it accordingly. Similarly, when presented with an unknown dog picture, our model can distinguish floppy ears from wet noses and wagging tails.

Deep learning (DL) falls under machine learning techniques inspired by human brain structure and function. Rather than relying on manually crafted differentiators (like "pointy ears"), deep learning models train themselves to discern characteristic features independently via layers upon layers of interconnected neurons designed for processing ever-increasing complex data sets. For example, building facial recognition models involves feeding thousands of images into your program while labeling each with its corresponding person's name. Labeled images facilitate identification by a deep-learning system of unique features that define individual faces like eye shape or nose-width or lip-curve. With time, this ability becomes more nuanced enabling recognition even when presented with previously unseen facial cues or configurations.

Artificial Intelligence (AI) represents various applications ranging from machine-learning strategies like deep-learning models fused with other methods-aimed at delivering smart technologies capable of reasoning or perceiving, just like humans. An instance of such is building a text-translation model where several thousands of examples in both languages have their correct translations labeled as guides for the AI to learn syntax and grammar rules pivotal for accurate translations. Over time, one could anticipate an augmented performance from the translation model as it fine-tunes its abilities to comprehend subtle nuances from different languages and dialects.

Machine learning, deep learning, and AI operate differently than conventional software in several significant respects. Particularly, their capability to learn using data while forgoing strict regulations that are defined beforehand. The result is a dynamic adaptability, which gets refined with novel experiences when encountering new situations or unforeseen roadblocks along the way. Moreover, unlike traditional software programs designed within strict systemic boundaries, AI thrives on complexity. With their capability to handle complexity in an adaptable manner, it's no wonder that machine learning, deep learning and AI are so highly regarded.

By leveraging vast amounts of data sources to identify patterns through pattern recognition technology, these systems offer a broader set of insights leading toward more precise decision making processes than ever before seen in technology applications. Besides being scalable against vast amounts of information, they have solved many issues where traditional software applications challenged by high volumes have struggled in the past, such as image classification or natural language processing tasks. This makes them indispensable when facing complex challenges not typically handled by existing systems.

Now, let us dive into the compelling narrative surrounding the history behind these fascinating technological breakthroughs. Artificial intelligence has made significant strides since it first emerged back in the 1950s. At that time, computer scientist John McCarthy introduced the term "artificial intelligence" to a world that was both skeptical and wary of machines being capable of human-like thinking and learning abilities. Despite the initial skepticism, McCarthy persisted in developing AI alongside his fellow researchers gradually bringing their aspirations to fruition.

One early success was the creation of Boris—an experimental chess playing computer program designed by programmer Dietrich Prinz in 1956. Although Boris had its fair share of challenges, it demonstrated how machines could use logic and reasoning to perform simple tasks.

As time passed, AI continued to grow, expanding its capabilities and during the 1980s, expert systems emerged as an innovative approach, which simulated human decision making using sophisticated rules and logic modeling. Leading experts have made remarkable advancements in AI technology over time that have brought about revolutionary changes in the field. One such incredible feat is Edward Shortliffe's MYCIN system designed to diagnose bacterial infections accurately and recommend treatment solutions. With significant progress made over time due to innovative developments within neural networks and machine learning systems like Deep Blue that defeated chess champion Garry Kasparov in 1997, modern-day society benefits from technology like self-driving cars and personalized advertising that utilize AI solutions.

The transformations expected from future innovations are vast—leading us on an exciting journey through an intriguing history of artificial intelligence. At one time, the concept of machines capable of replicating human thought processes seemed unattainable—an idea only found within sci-fi novels or movies. However, this did not deter McCarthy and fellow researchers from pursuing this ambition without fail. This ambition led them to host the Dartmouth Conference in 1956, marking a pivotal moment for AI research. Here, notable scholars such as McCarthy himself along with Marvin Minsky and Claude Shannon came together to explore how AI could be applied while dissecting its challenges.

Artificial Intelligence - First steps

One of the most significant stories from the early days of AI is the development of Boris, which was mentioned above, a chess playing computer program created in 1956 by programmer Dietrich Prinz. While Boris could only play a limited game of chess using a series of pre-programmed moves, it was still an accomplishment as it was one of the first machines that could perform a complex task using logic and reasoning.

In the 1960s, researchers expanded AI by creating programs like ELIZA, which was developed by Joseph Weizenbaum at MIT. The natural language processing program could simulate basic conversations between humans and computers even if it couldn't engage in true understanding of the conversation taking place. During the 1970s, AI research shifted to focus on expert systems, which are computer programs designed to replicate human decision making processes in a particular field. One notable example is MYCIN by Edward Shortliffe that could detect bacterial infections and suggest treatments. By doing this successfully, it established medical AI as a field with great potential. In the 1980s, neural networks became popular due to their capability to process data like human brains do. Frank Rosenblatt's famous Perceptron model, for instance, was an early example of a neural network that featured interconnected nodes responsible for data processing before transmission to the next layer. Despite its deep limitations in terms of functionality—being a scaled back model at the time—it was successful at demonstrating its potential as an AI tool.

In the 1990s and thanks in part to breakthroughs in machine learning and neural networks, developments in AI's popularity were surging again. AI utilizes Machine Learning techniques that enable machines to learn explicitly from data gathering without any explicit programming input. Machine learning advancements are well evident within several exciting fields like natural language processing and speech recognition and image identification tasks. As an instance, machine learning integrated into the Deep Blue chess playing robot facilitated its famous feat of defeating chess world champion Garry Kasparov in 1997.

AI development hasn't slowed down since then, as technology continues to adapt itself toward new applications and industries, enabling self-driving cars or personalized advertising features that promise to revolutionize how we work or interact.

In the past few decades, deep learning has been a significant point of AI progress with its subset of techniques involving integrating layers upon neural networks architectures, resulting in improvements across areas such as image recognition, natural language processing and speech recognition. The world of AI has had a groundbreaking victory recently when AlphaGo developed through Google's DeepMind used deep learning techniques successfully to defeat a High-Difficulty game Go World champion.

Additionally, Robotics applications have been impacted significantly by deep learning techniques implementation, such as Spot robots' creation by Boston Dynamics, which presented the capability of maneuvering through challenging terrain while performing tasks like carrying heavy objects and opening doors effortlessly.

Healthcare solutions providers today have turned to AI as an enabler for more personalized services provision. For instance, AI algorithms provide quicker cancer detection potential on medical images analysis, and enabling patients to get mental health counseling assistance provided via Ai-powered chatbots easily accessible any time of day.

The rapid forward progress is not without some concern. Widespread allegations about the impact on society continues to grow as AI evolves more seamlessly. AI's growing abilities have led experts to highlight potential job displacement risks along with privacy matters that arise with extensive data analysis by algorithms. Even so, civilization can expect considerable benefits from further technological innovation that will enable revolutionary transformations across multiple domains including work practices or social interaction modes at large.

The impact of Artificial intelligence on our lives has been present in the general and public domain for a few years. It has fascinated many creators within popular culture from science fiction stories such as those of Kubrick's classic, '2001: A Space Odyssey' and Cameron's, 'Terminator' franchise through to more current Netflix blockbusters like HBO's 'Westworld' series and Garland's screenplay 'Ex- Machina.' While earlier depictions often attempted to show human-like machines either turning hostile toward their creators (HAL) or poised against humanity all along (the Terminator), later works encourage contemplation over points such as ethics within creation or perception subjects like consciousness emergence and machine-human ties. A young programmer in the movie is invited by a tech mogul to assist in testing the intellect of Ava—his robot creation residing within a private estate. As he interacts with Ava further during his stay there, doubts arise over her ability to possess sincere feelings alongside true agency.

On another note regarding artificial intelligence portrayals within pop culture lies the movie *Her*, which provides insight into what it means for the protagonist who embraces an AI entity as their partner capable of not only catering but also fulfilling their deepest emotional needs.

Finally, evident in the literature standpoint is Neuromancer, which depicts AIs governing every aspect of life within dystopian societies. The portrayals relating to AIs reveal society's conflicting stances on this ever-evolving technology. Both hopeful for its potential and anxious of the implications of exceeding human intelligence, audiences are exposed to mature themes including ethics within the creation of intelligent machines, human relationships in the age of technology as well as dangers associated with over-dependence on automation. As a result, pop culture's AI narratives played a massive part in shaping our perceptions regarding this technology and how it'll impact us increasingly over time.

Neuroscience and AI

Although seemingly unrelated at first glance—with neuroscience centered on biological systems while AI focused on artificial ones—the overlap between these fields is significant. For many years, experts in both domains have been intrigued by how they may inform each other's work. One clear link is found through neural networks that vastly connect neuroscience with AI. Neural networks, which are inspired by brain structure and function, rely on layers of nodes or neurons that compute simple operations on their inputs. Then, they transmit results to subsequent layers while modifying the strength of node connections until they reach the capability to perform complex tasks such as identifying images, interpreting language or playing games impressively well. First proposed as a brain model several decades ago, neural networks have become integral components within both neuroscience and AI research. Both have common ground in their use of neural networks to model real neurons' conduct while trying to understand how these networks lead up to more complex human behaviors. On the other hand, AI utilizes these same network structures in building intelligent systems capable of learning from previous data and performing tasks thought to be exclusively human.

Learning is another shared area of concern between neuroscience and AI with researchers determinedly striving toward understanding mechanisms behind system progression through experiences resulting in improved functionality over time. Neuroscience researchers study neural plasticity's impact on the brain resulting from experience, while AI extensively employs learning algorithms for duties such as recognizing patterns within datasets or decision-making games.

By focusing on biology principles and replicating natural processes accurately within technological platforms, researchers are currently developing neuromorphic computing—an emergent subfield within artificial intelligence dedicated solely toward creating more biologically compatible artificial neural networks. To enhance machine learning models' agility, resilience, and adaptability beyond conventional techniques, researchers are developing systems that imitate structural as well as functional aspects of human cognition.

Before exploring further intersections, we must first ascertain clear-cut definitions for both Neuroscience and Artificial Intelligence disciplines themselves. Neuroscience is an all-encompassing multidisciplinary scientific field shedding light on nervous systems spanning from molecular biology to cognitive psychology, whereas Artificial Intelligence is concerned with creating intelligent machines capable of demonstrating human-like abilities such as understanding natural language processing or distinguishing objects on screen using visual perception. The point at which neurons interact with neural networks is one of the heavily researched challenges of neuroscience while Artificial Intelligence strives toward creating intelligent computer systems capable of performing tasks requiring human-like intellect. This is done by researching various methods including rule-based settings or machine-learning algorithms besides neural networks.

Despite their distinctive characterizations as separate fields, connections do exist between them. On one hand both neuroscience and AI aim at comprehending the essence of intelligence within intricate mechanisms and on the other hand, both try enhancing system performance levels by employing learning mechanisms based on what has been experienced before.

Undoubtedly, one of the most significant aspects shared by both neuroscience and AI is neural networks. In terms of neuroscience specifically, these play a foundational role as basic building units responsible for sensory processes all through decision-making complexities. On the other hand, its use within AI has led to creation of intelligent systems capable of learning from data coupled with various task performance including natural language processing or image recognition.

Looking back in history reveals how far we've come with our understanding of brains and nervous system mechanics. From ancient Greek fascination with this subject matter down to microscope invention during the 17th century enabling scientists to scrutinize brain structure exhaustively. In subsequent years, further discoveries included neuron identification as basic building units for the central nervous system, plus specialization emphasis based on observed regions within our brains. Modern technologies like electroencephalography (EEG) came a few decades later, which allowed measuring electrical activity in human brains while positron emission tomography (PET) gave researchers the ability to visually observe real-time brain functions.

The advancement of technology throughout history has reached a peak today wherein scientists have made significant progress understanding how our brain works—from the molecular details up until managing large-scale neural functions.

On another note, lies the rising field of Artificial Intelligence and Artificial Neural Networks posing as a prolonged fascination ever since its emergence when researchers started exploring ways to create machines capable enough for tasks that otherwise require human intelligence—which dates back in the early years of the post-WWII era. Early attempts focused more on rule-based models employing if-then statements; however, limitations were prominent on handling complex or more ambiguous information sets.

It wasn't until the later decades (in the 1960s – '70s), that researchers introduced better approaches for AI development such as machine learning or neural networks, offering advantages such as learning patterns from given data while applying probabilistic reasoning instead of sticking to rigid rules. It took time but eventually, in the past decades (1990s – '2000s), groundbreaking milestones for AI were achieved and developed that established its modern-day framework. Neuroscience and AI also found a significant link today with each field benefiting from shared insights and techniques. Advancements in neuroscience have led to exciting progress in AI, specifically with deep learning techniques utilizing Artificial Neural Networks modeled after our brains. As such, innovations continue to evolve, scientists can now simulate and model intricate neural networks while also analyzing large scale datasets for a deeper understanding of the brain.

The pursuit of understanding how our brains work traces back thousands of years amongst scholars who sought answers through early philosophical exploration and scientific endeavors. The microscope's introduction during the 17th century revolutionized research by providing unparalleled visibility into neuronal morphology and function. Santiago Ramón y Cajal pioneered this field further by introducing Golgi staining for detailed visualization on an individual neuron level. Today, his legacy as an essential figure within modern day neuroscience lives on. The contributions made by Ramón y Cajal were pivotal to our current understanding of how neurons work together within the nervous system. This visionary pioneer's work laid a solid foundation for future investigations into modern neuroscience research—so much so that he received a distinguished Nobel Prize in Physiology or Medicine back in 1906. While it took some time for this discipline to fully take on technological advances like EEGs and PET, scans helped catapult Neuroscience into a new era of discovery.

A key milestone during this period was the identification and subsequent study around action potentials, which ultimately paved new avenues for theories on neural function. The field of neuroscience made a significant advancement with its discovery of neurotransmitters—chemicals released by neurons that facilitate signal transmission between them. This breakthrough contributed to an insight into several neural functions including how substances like drugs affect our nervous system. Synesthesia captured attention among researchers during the late 19th and early 20th centuries; an intriguing phenomenon where stimulation of one sense triggers an experience in another sense entirely. Russian painter Wassily Kandinsky was among one of its most famous cases, experiencing colors as shapes and sounds as colors being just some examples that he reported feeling.

A thorough understanding of exactly how our brains process sensory information was studied intently by scientists like Sir Francis Galton and Richard Cytowic through their investigations into synesthesia research. Neurosurgeon Wilder Penfield paved new pathways for brain mapping with his inventive use of electrical stimulation during exploration efforts back in the 1930s to 1940s era. As a renowned neurosurgeon, Penfield utilized a unique approach during surgeries of epilepsy patients by using electrical probes to stimulate various areas of their brains while they were fully alert. This method allowed him to determine precisely which regions were responsible for specific functions such as movement or language.

Throughout mid twentieth century America, Roger Sperry—expanding upon Penfield's techniques—conducted research on individuals who had undergone procedures severing their corpus callosums which joined both hemispheres. Sperry determined that one side was specialized for analytical thinking and language processing while creativity emerged strongly over spatial reasoning ability from another side. In further studies during the seventies and eighties, Michael Gazzaniga delved even deeper into understanding how split brains operated by examining neurological responses to stimuli. His work added greater clarity about hemisphere specialization among patients who had undergone operations severing corpus callosum fibers between either side of the brain.

The idea that our brain hemispheres can operate independently with individual thoughts and decisions challenges long-standing beliefs about universal consciousness, as outlined in Gazzaniga's research findings. Scientists have been inspired by this work on self-awareness, leading to an explosion of growth in the neuroscience field today. Using varied techniques such as cell imaging or mapping neural network organization helps researchers study everything from molecular function to high-level operations.

Lastly, neuroscientific research inspires new innovations across multiple industries- such as Brain Computer Interfaces (BCI) technology granting individuals with disabilities control over devices using only their thoughts. With techniques such as EEG and functional magnetic resonance imaging (fMRI), BCI can detect patterns within brain activity and translate them into computer generated commands. AI researchers further this field by developing new methods to detect these patterns while simultaneously advancing other forms of brain machine interfaces. A result of this work is that scientists gain greater insights into neural function using simulations created through cutting edge AI technology—uncovering critical aspects concerning information processing while examining complex behavior. This newfound knowledge helps refine neuroscientific modeling while also contributing to the development of theories related to mind and consciousness.

The first steps in AI research began slightly after the cutting edge advancement in neuroscience research. Warren McCulloch and Walter Pitts introduced neural network models during computing's early days, which utilized artificial neurons for performing advanced logical computations. It has been well documented that interconnected neurons within networks possess extraordinary abilities such as complex computations featuring pattern recognition, decision-making among others. Frank Rosenblatt's groundbreaking research during the fifties through sixties culminated in his development of one of the first practical neural networks christened The Perceptron. This invention could learn experientially by adjusting the weights of its connections based on feedback while also recognizing data patterns.

Rosenblatt's work had far-reaching implications for artificial neural network analysis leading to the establishment of machine learning as a field. However, researchers shifted their focus away from neural networking between the seventies through eighties toward rule-based approaches for AI such as expert systems, which led to a slump in research on artificial neural networks. Nonetheless, crucial and necessary developments took place during the same decades, such as substantial improvements in computing power between the nineties onwards along with introduction of new algorithms like backpropagation. This resulted in the ability to train significantly larger and more complex neural networks. This renewal sparked interest once again and today these highly intricate structures play an integral role in many AI applications that encompass tasks ranging from image recognition to natural language processing alongside robotics.

While Artificial Intelligence has seen remarkable improvements thanks to neural network technology, criticisms have been raised over comparisons made between them and organic neurons within the human brain. Fundamentally, they share similarities regarding information processing but differ greatly where structure and functionality are concerned. Despite such critiques, research continues on both disciplines for insights into unlocking secrets held by various forms of natural neuronal activity while also opening doors toward exploring fresh fields of innovation within AI development.

The ability of neural networks to learn via experience was initially conceptualized by Hebb through his theoretical framework. Further advancements were made within this field thanks to contributions from Kunihiko Fukushima, a pioneering Japanese researcher who created the Neocognitron neural network during the 1980s using hierarchical processing methods for visual pattern recognition. Low level neurons detect simpler features such as edges and corners while higher level ones identify more intricate features like shapes and textures. This approach to neural networks encouraged researchers in the area of computer vision and stressed the significance of hierarchical processing.

In the 1990s, Yann LeCun made significant strides in image recognition by developing Convolutional Neural Networks (CNNs). The Convolutional Neural Network operates by applying a few filters onto an image that detect distinct features such as edges or textures. The output goes through several refining layers before ultimately classifying the image. This innovative idea introduced by LeCun has led to deep learning, which employs neural networks layered many times over to carry out intricate tasks including speech and image recognition.

Following LeCun's innovative research, Geoffrey Hinton made substantial contributions toward deep learning advancements when he introduced Deep Belief Network (DBN). DBNs are designed with unsupervised learning in mind where they can learn how best to represent statistical structures without direct guidance. Hinton's work on DBNs was revolutionary for deep learning advancements setting up many other popular architectures employed by AI applications today. During research on how our brains operate in relation to neuronic connections during the '80s, a scientific group known as 'connectionists' settled on a different approach founded upon intricate interconnectedness between individual neurons found within that system—hence having faith in it being shaped as one big neural network. Thus, commenced experiments involving what would be known today as 'artificial neural networks' or connectionist models of neural networks, with their mission being simulating neuronic network behaviors. As mentioned before, one of the earliest artificial neural network models was known as Perceptron and originated from America in the 1950s by a psychologist called Frank Rosenblatt. It developed simple pattern recognition abilities via artificial neuron connection adjustments. However, it could only recognize limited complex behavioral patterns compared to newer and more sophisticated-designed models like Parallel Distributed Processing (PDP) neural network.

The process of information processing entails relaying signals across different layers that each performs their own unique computations. With connectionist models, one key embedded attribute is the ability to learn through experiences gained from past interactions with data—whether supervision of supervision can dictate this data handling process.

Supervised learning processes, such as labeling examples adequately, trains the network on well labeled sets in order for it to predict future label output right. However, topics like unsupervised learning processes where only raw data sets exist without sufficient labeling leads network system developers to craft alternative methods that allow complex learning systems to learn statistical structures inherent within this uncertain dataset space.

Connectionist models have found diverse application in the real world outside of the academic domain, including but not limited to speech recognition, image recognition, natural language processing and game development. Deep Blue by IBM remains one historical example where connectionist model was used as automation tool to defeat world chess champion Garry Kasparov back in 1997.

Despite remarkable wins achieved using these models, criticism however has been leveled against their lack of biological realism in terms of not incorporating features such as glial cells and brain structure connections into their functioning models. There is no denying that, when it comes to understanding complex systems' behaviors, connectionist models have proven to be an invaluable instrument that has significantly influenced AIs development as we know it today. To put things in the right historical context, the earliest stages in creating these neural network models began around the late forties and fifties of the previous century, with researchers searching for inspiration from how our brains functioned. In those early steps of Artificial Inelegance, however, such initial designs remained quite rudimentary since they could only consist of just a few neurons along with simple connections between them at best. The available computational resources were simply too limited to support any incremental enlargement of those architectures. Despite the existing limitations operating at the time, the initial designs of artificial neural networks served as an essential cornerstone worth building upon later on.

Fast forward several decades later into the 1980s, where significant breakthroughs arose thanks to researchers who developed connectionist models—perhaps one technological feat that propelled AI toward even more astonishing heights than we ever thought possible before then. Such developments involved networks composed entirely of interconnected "neurons" or nodes; each one receiving input from other neural clusters while simultaneously transmitting output back to the overall network.

As time progressed, these models learned to adjust their respective "weights" using data that represented the connections between neurons. One such popular example of an early implementation for connectionist models was Frank Rosenblatt's Perceptron that was developed in 1950s America. The remarkable strides made within the discipline of machine learning are thanks to some extraordinary inventions over the years. In particular, Frank Rosenblatt deserves accolades for his outstanding creation— The Perceptron—back in 1957. This single layer neural network methodically analyzed object features and went on to become a catalyst for multi-layer networks' growth and development.

Geoffrey Hinton continued Rosenblatt's legacy with his brilliant invention known as The Boltzmann Machine—introduced decades later—which opened doors to understanding statistical data structure and uncovering hidden patterns within datasets. Hinton's groundbreaking work guided the development of deep learning—ensuring constant innovation and a world of possibilities. Another significant stride occurred when Yann LeCun introduced us to Convolutional Neural Networks in the early 1990s. The CNN played a critical role in image processing, allowing machines to recognize objects by breaking visual information down into several layers of abstraction. LeCun's seminar work on CNN in the field of computer vision has evolved into AI systems capable of detecting faces or identifying objects in real-time video streams.

Following these inventions, James McClelland introduced the Interactive Activation and Competition (IAC) model in collaboration with other psychologists during the mid-1990s. This model aimed at replicating human brain learning, using bottom-up top-down models for learning letters/words. This research has helped increase knowledge about our brain's visual processing, resulting in AI systems that can read and write texts or recognize handwriting. Andrew Ng, considered to be one of the founding fathers of modern AI, jointly with other computer scientist contemporaries advanced another connectionist model called Sparse Coding algorithm between the late 1990s-early 2000s.

By designing an algorithm that could learn to represent visual information efficiently through sparsity, we were introduced to an innovative solution; The Sparse Coding approach discovered highly informative image features that were vital for object recognition. Thanks to Ng's contributions which laid the foundation for deep learning's unsupervised methodologies, AI systems can now successfully recognize and track objects such as self-driving cars within real-world scenarios. These are just a few examples out of the many exciting advancements witnessed within connectionist models of neural networks over recent decades, with more progress undoubtedly set to emerge.

When comparing the advancements of AI and artificial neural networks with neuroscience, one should emphasize that at the very heart of neural plasticity lies the neurons' extraordinary ability to adapt their connectivity in response to diverse experiences encountered during a lifetime. This ability is termed by researchers as synaptic plasticity whereby connections between neurons either strengthen or weaken depending on various activity patterns. Extensive research conducted on animal models has facilitated identifying numerous critical molecular and cellular mechanisms responsible for synaptic plasticity. One noteworthy discovery in this field is that long term potentiation (LTP) significantly contributes toward enhancing learning capabilities and memory formation. By reinforcing synapses through repetitive activity patterns, LTP enables efficient signal transmission between neurons thus strengthening memory consolidation in the brain. Current research indicates that non-invasive brain stimulation techniques such as transcranial magnetic stimulation (TMS) or transcranial direct current stimulation (tDCS) could boost neural plasticity resulting in improved cognitive performance.

Several techniques can be used to promote neuroplasticity within the brain including applying a weak electrical or magnetic field. Cognitive training programs are also gaining traction as a promising avenue for enhancing specific mental abilities through practice exercises and tasks that offer feedback to users over time. This approach has shown favorable results in studies conducted among older adults where there was marked improvement not only within targeted areas but also overall cognition. Additionally, using implanted neural devices that stimulate neurons through electrodes could offer an innovative way of restoring lost or impaired cognitive function. While still early in development, this technology shows potential for treating a variety of neurological and cognitive disorders affecting different individuals' abilities negatively.

Overall, exploring neuroplasticity offers an exciting glimpse into how we learn while also presenting opportunities in developing new therapies and technologies aimed at enhancing our cognitive abilities. With ongoing research across these fields, better health outcomes are possible sooner rather than later for those struggling with neurological issues affecting their quality of life negatively.

Just like learning models based on artificial neurons, neural plasticity highlights the brain's ability to adapt over time due to experiences or environmental stimuli resulting from modifying the strength and connectivity patterns of synapses between neurons. Synaptic plasticity is a well-researched form of neural plasticity that explores how synapses strengthen or weaken due to altered activity patterns.

In contrast, learning constitutes new information acquisition reshaping neuron activity patterns along with modifying connection strengths for encoding purposes resulting in synaptic connection alteration closely related with neuroplasticity principles. Both processes require extensive research covering complex mechanisms relying on molecular reactions like neurotransmitter level alterations; intracellular signal pathways activation; gene expression modulation besides cell based structural changes at a network level.

Although traditionally studied using animals, cutting edge technology such as neuroimaging and neurophysiology have facilitated direct research into such processes within humans. Recent studies into neural plasticity have provided a more profound comprehension regarding human cognitions underlying neural mechanisms. As a result, these findings have led to novel methods for enhancing cognitive performance treating different neurological disorders effectively.

Neural plasticity remains one of the salient areas of neuroscience, where it intersects with Artificial Intelligence since analyzing cognitions mechanisms contributes toward developing new AI based algorithms.

Cognitive research roots, developed in separate but in parallel to traditional neuroscience which we have already laid out, date back to scientific research conducted by pioneers like Ivan Pavlov and Edward Thorndike during the early 1900s. Their work greatly impacted early theories on classical conditioning. Pavlov famously discovered how neutral stimuli could become associated with reflexive responses when paired repeatedly with unconditioned stimuli while Thorndike developed "the law of effect"—suggesting behaviors followed by positive consequences that repeat more often.

During the mid-20th century, researchers like Donald Hebb and Eric Kandel further expanded on these ideas contributing vital insights into neural plasticity mechanisms and paving the way for more groundbreaking research to come. Neuroscience owes allegiance to two great researchers: Donald O. Hebb, who came up with Hebbian Learning Theory; Eric Kandel whose study focused predominantly on Aplysia sea slugs regarding the neural basis for memories formation. He demonstrated conclusively that changes in synaptic strength lead to long-term memory retention.

As we moved into modern times, innovations such as electrophysiology imaging technology allowed scientists to observe accurate readings of brain functions. These tools also brought clarity into our deep understanding about neural plasticity and learning mechanisms such as Long term potentiation and Long term depression, responsible for strengthening or weakening synaptic connections.

In the present day, researchers continue to examine how neural plasticity and learning shape our thoughts regarding aging, development, and diseases such as neurological and psychiatric disorders. In addition, stimulating brain tissues using non-invasive techniques alongside cognitive training protocols is vital in enhancing memory retention in people with intellectual disabilities. The fascinating field of neural plasticity has expanded our comprehension regarding memory and learning mechanisms dramatically. In addition to helping treat cognitive disorders profoundly better, this knowledge can aid in developing sophisticated machine learning algorithms or complex neural network models imitating biological processes more accurately. In one remarkable research experiment conducted by University College of London, the brain structures of bus drivers and licensed London taxi drivers navigating without GPS were scrutinized. It was discovered that taxi drivers had significantly larger posterior hippocampus, i.e. spatial navigation and memory control centers indicating how experience or practice can change the brain structure via neuroplasticity.

Further to this undoubtedly fascinating insight, the National Institute of Mental Health scientists carried out a study observing how monkeys responded to different objects in the 80s. Jody's story from the 1980s is a powerful example of neural plasticity. After receiving a corneal transplant that restored her vision, she faced challenges interpreting the visual information she received. But with time, her brain adapted and changed significantly to accommodate this new input.

Influences from the environment can significantly impact neural plasticity and learning processes. This finding could prove crucial for educational or therapeutic purposes. Moreover, studies revolving around this topic have strong connections with developments in artificial intelligence research; particularly in the case of training neural networks that imitate biological analogs.

Typically, artificial neurons within these networks undergo adjustments in connection strength in order to recognize patterns or predict outcomes—just as our own brains change through experience and learning responses. In order for this process to occur accurately when training a network, labeled examples (e.g. cat versus dog images) need to be present so that adjustments can be made until outputs are improved gradually—a process known as "training" the network. In the case where new inputs arise, in some way different than the original labels upon which the model was initially trained on, or if changes occur within its environment, we need to maintain similar accuracy levels of the model's retraining requirement.

Similarly, *Adaptation* is a pivotal aspect of neural network design referring to the ability of such systems to adjust and fine tune their responses when confronted with new sources of input. In order to create neural networks with greater flexibility, efficiency and adaptability researchers are looking toward principles of neural plasticity for inspiration. One example of this trend involves exploring the use of *spiking* neural networks, which mimic bursts of electrical activity like those seen in biological neurons. Such systems may be especially helpful in temporal processing tasks such as speech recognition.

Cognitive neuroscience lies at the intersection of scientific disciplines studying the cognitive processes that inform our everyday experience: perception, attention, memory, language and decision making.

AI development, at least at the early stages of this domain, utilizes heavily related concepts from the various fields of study in neuroscience.

The Challenges of AI Creation

Developing learning algorithms is a tricky business. There is a plethora of obstacles that one must overcome to create an algorithm that can learn from data without overfitting or underfitting, while dealing with noisy or incomplete data. To add to this, there is the curse of dimensionality and the issue of biased data. These are just some of the many challenges that make developing learning algorithms a daunting task.

To illustrate the challenge of overfitting, let's consider a story from a group of researchers who were trying to develop an algorithm that could identify patients with depression based on their speech patterns. They trained their algorithm on speech samples from depressed and non-depressed patients, but the algorithm was performing poorly. They discovered that the algorithm had memorized the specific vocal characteristics of the speakers in the training data, and was not able to generalize to new speakers.

In another example, a group of researchers were developing an algorithm that could identify signs of cancer in medical images. They trained their algorithm on a large dataset of images, but found that it was only accurate about 50% of the time. They discovered that some of the images in the dataset had been mislabeled, causing the algorithm to learn the wrong features and perform poorly on new data.

The curse of dimensionality can be a significant challenge in developing learning algorithms. To illustrate, imagine trying to classify images of objects that are made up of a large number of small pixels. Each pixel adds a new dimension to the data, which can lead to the so-called *curse of dimensionality* where the amount of data required to learn the underlying patterns, grows exponentially.

Finally, the issue of biased data is one of the most significant challenges in developing learning algorithms. Algorithms can be trained on biased data, leading to biased results. For example, a team of researchers developed a facial recognition algorithm that was trained on images of mostly white men. The algorithm had poor accuracy when it came to identifying women and people of color because it had not been trained on enough diverse faces.

Developing learning algorithms can be a challenging and complex process that requires a lot of work to get right. Compared to other technologies, these algorithms can be difficult to develop due to several factors.

One major challenge is the lack of transparency in how these algorithms make decisions. Unlike traditional software systems where the rules and logic are explicitly programmed, learning algorithms rely on patterns in data to arrive at their decisions. This can make it difficult to understand the precise reasons a certain decision was made or identify potential biases in the data.

Another challenge is evaluating the performance of machine learning algorithms. It's not always clear what the correct answer is, especially in cases like image recognition and natural language processing, where there may be multiple correct labels for an image. This can make it tough to evaluate the accuracy of the algorithm thus choosing the correct solution for a given problem that might not be straight forward.

Balancing accuracy with interpretability is another challenge in developing learning algorithms. Highly accurate algorithms may be very complex and difficult to understand, making it tough to identify and correct errors or biases. Additionally, learning algorithms often require large amounts of labeled data to train effectively. This can be a challenge in cases where labeled data is scarce or expensive to obtain.

Lastly, machine learning algorithms can be less flexible than traditional software systems when it comes to adapting to new data or changing circumstances. They may need to be retrained from scratch when new data becomes available or when the underlying distribution of data changes. Even the slightest skew from the original dataset used to train the algorithm initially, could lead to the devastating decrease in the model's performance.

All these factors can make developing learning algorithms a challenging and nuanced task. They are conceptual in nature, so that despite their presence in virtually any learning model solution, harnessing careful considerations and attention to them enables creating effective and impactful learning systems. So, as we delve deeper into the world of AI, it's essential to keep these challenges in mind to develop robust learning algorithms.

In addition to the conceptual challenges mentioned earlier, practical limitations such as coding and debugging learning algorithms can also be difficult. These algorithms can be highly complex, often involving multiple layers of interconnected neurons that are trained to recognize patterns in data. As a result, even small coding errors can have significant impacts on the performance of the algorithm.

Debugging these algorithms can be particularly challenging because the decisions made by the algorithm can seem opaque or unintuitive. A seemingly minor change in the code can cause the algorithm to misclassify a significant number of images, and the reason behind the misclassification may not be immediately clear.

To make matters more complicated, debugging deep learning models often involves dealing with vanishing gradients and exploding gradients, which can cause the training process to stall or diverge. These issues can be difficult to detect and fix, requiring in-depth knowledge of the underlying mathematical concepts.

To help address these issues, developers often rely on visualization tools to help them understand the behavior of their algorithms. These tools can provide visual representations of the inner workings of the algorithm, helping developers identify and fix issues more efficiently.

Despite the challenges, the rewards of successfully developing and debugging a learning algorithm can be significant. By leveraging machine learning algorithms, it's possible to create systems that can learn from data and make intelligent decisions, opening the possibilities of immense breakthroughs in fields like healthcare, finance, and transportation.

MACHINE LEARNING

n this chapter, we embark on an epic journey, deep-diving into fundamental concepts embracing its diverse implications—investigating critical issues and challenges while optimistically examining emerging opportunities within a flourishing domain rich with creativity.

Machine learning encompasses advanced Artificial Intelligence technologies enabling computers to learn from data without resorting to explicit programming but using sophisticated methodologies instead. Recognizing underlying patterns within large swathes is crucial for machine learning systems as this helps them generate insightful results such as predictions, discovering unusual events or identifying novel discoveries.

We will commence the chapter's comprehensive overview with exploring different types of machine learning methods, dwelling separately on supervised and unsupervised techniques.

Supervised learning employs labeled data for predicting new outcomes while unsupervised learning focuses on identifying the patterns found within unclassified or unlabeled data. Our journey continues by investigating fundamental building blocks like regression and classification, which are vital in developing several machine learning algorithms garnering importance in various industry domains. Regression techniques routinely predict continuous values, stock prices, for instance, while classifications assess inputs to specific categories like spam emails.

Our exploration traverses through some of the most common applications of Machine Learning being leveraged currently—encompassing sectors like natural language processing (NLP), image recognition as well as cover fraud detection and predictive maintenance concepts pivotal to industries including healthcare, marketing, manufacturing and finance which have already been transformed positively.

However promising the applications may seem to us, quite inevitably they raise potential ethical concerns and algorithmic bias implications when insufficient datasets without neutrality or sufficient diversity are used, rendering them unfit for their designated use case intended. Therefore, this heightens the challenges associated with creating accurate algorithms that are aimed at accurate representations of diverse samples across varied cultures globally.

There's a continued need for growth in this industry, both in tools for development as well as finer understanding of the mathematical foundations. Hence, we need to view the continued development toward emerging trends critically when dealing with this ever-evolving landscape characterized by fast-paced innovations, which makes data usage accessible across diverse industries, offering tangible rewards.

Through our journey's culmination after rigorous immersion with machine learning principles unraveling adaptive ways it can be implemented—you'll become equipped with fundamental skills imperative in developing a firm grounding within multiple industries and contexts. This will help sensibly guide best practices and promote meritocracy.

Let's take a paradigm shift into an exciting world of limitless potential. Welcome aboard!

At its core, Machine Learning is a process in which computers can learn from data without explicitly being programmed to do so. On paper, this means that as we expose these systems to a greater pool of information, they become better at performing their tasks. However, let's delve deeper into the subject in hand by breaking it down into more relatable human terms.

If we take teaching a computer to recognize photos of cats as an example, initially we could attempt manually programming it with features that are characteristic to this species such as whiskers and furry tails. Due to the nature of cats, this is heavily time-consuming and indeed not highly precise or sophisticated. Instead of considering the classical approach of solving complex problems, we can present the learning models with many images of felines that share certain traits with each other until they get a good understanding of the characteristics themselves, aiding the algorithms in identifying similar pictures within its internal database.

While the above description is an over generalization of what machine learning is, it is common to classify the Artificial Intelligence domain into two primary forms of learning: supervised learning and unsupervised learning.

As mentioned earlier, to supervise learning, we have to introduce computers to labeled data wherein it understands what the correct formulas should be for specific inputs given to them. For example, in cat image recognition when using supervised learning, we may choose to label images with cats in them *cat*, whilst tagging those without cats as *not cat*. The machine eventually learns how identifying features interlock with the image label *cat* through data trial and error, and learning from them in a non-random manner. To accomplish this we have to repeatedly exhibit learning models with examples representing the target distribution. It is done so that machines gradually recognize emissions when compared with correct outcomes achieved when correctly evaluating all potential test imagery, hopefully based on the same distribution from which the training dataset was extracted.

Supervised learning is a way to teach computers to do things by showing them a high number of examples. It is similar to teaching a child how to do something by showing them how to do it over and over again.

Let's say, you want to teach a computer how to recognize pictures of dogs. You would start by showing the computer a great number of pictures containing dogs and telling it, "Hey, these are all pictures of dogs." This is like showing a child pictures of dogs and telling them, "Hey, these are all pictures of dogs."

Once the computer has seen enough pictures of dogs, it can start to make predictions on its own. So, if you show the computer a new picture of a dog that it hasn't seen before, it can say "Hey, that looks like a dog to me!" This is reminiscent of a child looking at a new picture of a dog and saying, "Hey, that's a dog!"

A key example illustrating supervised learning's potential is image recognition technology where high-quality machine datasets are employed identifying unique characteristics to make accurate predictions about the labeling requirements of novel images quickly.

Another notable supervised learning category stems from leveraging historical customers' preferences and trends over time to obtain always much-needed insights useful in predictive analytics across various domains.

As such, utilizing supervision in analyzing data can foster better decision-making efforts while catering to predictive use cases' needs uniquely but requires rigorous adherence toward larger datasets while avoiding inherent biases building up over time during organization processes and routines.

In unsupervised learning, we don't provide the computer with labeled data. Instead, we give it a bunch of data and let it find patterns on its own. Going back to our cat example, we might give the computer a bunch of pictures of cats and dogs without telling it which ones are which. The computer would then group the pictures into separate clusters based on the features it finds.

Unsupervised learning is another way that computers can learn things, but instead of being shown examples like in supervised learning, the computer has to figure things out on its own, without any guidance. Unsupervised learning is undoubtedly one of the most remarkable forms of machine learning that enables computers to identify underlying patterns and insights in data that humans would often overlook otherwise. It mirrors the work of a skilled detective who can accurately analyze a set of seemingly unrelated clues and unveil hidden connections between them.

Think of it like a child exploring a new environment on their own. They don't have anyone telling them what to do or showing them what's important—they have to figure it out for themselves.

In both of these available approaches, model training stage enables us to choose appropriate algorithms for optimal performance evaluating models based on unseen new datasets. After successful completion of training and assessment phases, deploying the model becomes a feasible a part of an actual application's development efforts. To achieve this objective, it may be necessary to integrate newly developed models into current software infrastructure or employ them in making decisions/predictions instantly.

In one use case example for unsupervised learning manipulation—image clustering algorithms—the computer groups similar images together without any pre-established labels that could be helpful in organizing vast image datasets and identifying essential information patterns.

Another example pertains to anomaly detection, one where rare or unusual events within an established dataset can be identified through unsupervised machine-learning methods. Suppose we have credit card transactional data that falls outside their regular behavioral patterns. In that case, we can utilize this method to detect suspicious transactions on the internet or any digital playground for purchasing goods—representing potential fraud occurrences among other malicious activities.

Similarly, market segmentation is another shared application area for businesses, for grouping customers based on particular behavioral preferences, in order to get insightful customer-base understanding, which enables tailor-made marketing strategies accordingly.

On another note, topic modeling represents an intelligent solution using unsupervised learning algorithms capable of analyzing large text-data collections by pinpointing primary themes or topics like editorial quality control or news coverage trends within relevant news cycles or media types.

As discussed earlier, unsupervised machine-learning possesses incredible efficacy in discovering significant datasets overlooked by conventional analysis methods at its core. The unpredictability of today's methods has made it more challenging than supervised techniques relying entirely upon explicit guidance or supervision. Supervised approaches hold many benefits, albeit while attempting to bridge the gaps pertinent with its unpredictable counterpart's failings but hold enormous potential for breakthroughs across various administrative sectors like fraud management, regulation support and others.

To sum up briefly—Unsupervised Learning provides numerous opportunities promoting dynamism within the rapidly-evolving technological landscape while not relying on explicit guidance that human-read metrics may, creating unique opportunities across various fields and applications where traditional analytics may fail.

In our contemporary society, we can barely avoid encountering data in whichever sector we operate. From companies collecting masses of clients' information down to researchers' probing experiments for insights; how do we make meaning out of all this magnitude? Well, look no further than machine learning algorithms. They serve as an excellent set of rules that a computer adopts to learn from gathered information and make predictions eventually.

Most commonly used methodologies among the field of machine learning include:

Linear Regression Algorithm: popularly utilized for predicting continuous value instances such as home prices predicated on size footages represented by drawing straight lines through existing data points; thereby producing probability prediction models for future events happening from current unto thereof. Linear regression remains one of the most powerful tools across various fields like economics, medicine and engineering when it comes to predicting necessary values based on specific data points through gaining insight into variability between established factors and vectors.

As opposed to Linear regression, Logistic Regression offers us predictive capabilities galore when it comes down to earmarking a chosen event's chance or likelihood—facilitating business research well. The usage of this approach could be, for example, promoting sub-modalities within recurring client purchase behavior toward certain product selection respectively. Classifying incoming content effectively works by utilizing 'true/false', 'spam-not spam' labeling schemes leveraging algorithms-driven technology. Logistic Regression implemented via measured key factors gives us exact results derived from applicative machine learning concepts based on metadata and messaging source account id's usage patterns, illuminating vital signals amongst ever-changing signals online.

Researchers commonly apply logistics regression techniques when working with probabilistic classifications derived from datasets involving multiple variables. In healthcare, anomalies like susceptibility variations toward medical conditions manifest themselves differently among individuals. Hence, logistic reconstruction techniques support these evaluations with enhanced levels of accuracy. For example, one implementation involves using patient-specific medical records alongside detailed physical characteristics like BMI scores for assessing the likelihoods of their acquiring potential illnesses.

In marketing context, fundamentals such as customer purchase histories are used in conjunction with additional data points like demographics or geographic information to develop predictive models. Merchandisers use logistic regression in these solutions to make informed decisions on what products would perform best depending on the characteristics of the customers in specific locales.

Likewise, electoral research efforts increasingly rely on logistic regression aided models for assessing voter inclinations and the resulting political race outcomes. The techniques deployed account for varying factors like polling data ratios, past election results trends and local circumstances yielding more analytical reliability.

Overall, this approach has proven to be a reliable method across diverse subject areas, offering increased predictability based upon probability theory insights. If you're deciding what outfit would be suitable for today's weather conditions – hot or cold – an analytical approach would be asking yourself pertinent questions before making moves concerning clothing choices. For instance: "What clothes can I wear when it's scorching [or chilly] out here?"

Likewise, Machine Learning model building using Decision Trees stem from asking certain questions and making decisions rooted primarily on its informative features derived from new input data sets [training datasets analyzed initially]. Similar to fashion choices based solely upon weather patterns, when we try to develop an algorithmic solution within ML for Predictive Analytics that is concerned with estimating consumer behavior patterns, customer interests, we might isolate customer spending patterns by taking into account variables such as age range, income bracket, purchasing habits from previous shopping history. These will inform our specific rules in our decision trees. Starting with the root node of asking whether someone [potential customer] is over 30 years old amongst other such tests precipitating branches represented by outcomes for each question/answer used to further refine potential customers for sales leads, best candidates for promotions, etc. In machine learning applications, a randomly generated forest of decision trees is utilized to increase precision levels when predicting outcomes based on input data. Contrary to relying upon one solitary tree, the approach overcomes possible overfitting by drawing upon numerous trees. This approach involves multiple subsets of diverse input data features from all available information on hand with respect to retaining diversity amongst them. This in turn helps attain improved algorithm performance for more accurate outcomes in its final outcome prediction when predictions are made or conclusions drawn from said dataset.

In machine learning, K Nearest Neighbors (KNN) is a model utilized in both classification and regression tasks that depend on the identification of up to K closest data points from input data relevant to accurately predicting/classifying occurrences using either majority classification or average percentage utilized. An example of application of this model would involve predicting whether customers are likely to purchase your product based on similarities between them and other recent customers who've purchased it. Similarly, when used by streaming services like Netflix etc., KNN recommends various movies or series depending on viewers' preferences and similar user choices. Let's consider an instance where we apply this analysis for predicting whether a fruit such as apple or orange should be classified based on its weight and texture. When using KNN for these predictions, we search for the nearest data points relative to our fruit based on weight and texture from within a predefined cluster size preferably set at equal distances/K nearest neighbors. If most among those closest clusters represent oranges, we predict an orange as our target category; otherwise we go with apples. While straightforward in usage, accurate selection of output variable values must be practiced concerning appropriate normalization techniques across use case scenarios inferred before initializing this tool. The assortment provided by available tools within machine learning extends beyond those outlined above. The decision-making process when we select an appropriate procedure relies upon established expertise within respective fields, along with an understanding between corresponding datasets and end-results desired. Gaining familiarity across this diverse array allows us to translate unmanageable quantities into something more accessible via deployment in analytics purposes whilst ensuring proper comprehension in-house boundaries exists before attempting analysis projects.

Everyday Use of Machine Learning Algorithms

Machine learning algorithms have gained immense popularity in our day to day lives, playing a crucial role in everything from recommending movies on Netflix to suggesting new products on Amazon. Utilizing relatively simple models, such as the ones mentioned earlier, can be fairly useful for a variety of applications and fields. Let us present a few instances below.

Linear Regression: Predicting Home Prices

If you're searching for your dream home but cannot go beyond your budget, linear regression can come to your rescue. Linear regression is a machine learning algorithm that creates a line to fit data points best and uses it to predict new values, such as the price of a house considering its square footage, location, number of rooms and amenities.

Logistic Regression: Making Informed Choices

Logistic regression is another machine learning algorithm used to predict the likelihood of an event taking place with precision. Logistic regression assists businesses by providing insights into customer behavior and predicting their purchasing probability based on relevant factors such as age or health status. Based on these predictions, companies can modify their marketing strategies and enhance product development. Likewise, insurance firms use logistic regressions for determining risk factors while insuring customers.

Decision Trees: Personalizing Your Shopping Experience

Have you ever wondered why online shopping websites suggest items you might like based on your browsing history? By using decision trees for classification through an array of yes or no queries based on age, income or location, providers gauge whether a customer is inclined toward purchasing a particular product or not. With this information available, retailers can customize shopping experiences personalized effectively.

Random Forest: Enhancing Medical Diagnostics

Random forest comes under the same category as decision trees but uses multiple trees for making predictions rather than just one. In medical diagnosis context, Random Forest has been essential, given its ability to determine connections between various conditions. It makes accuracy more reliable when it comes to diagnosing healthcare related issues giving necessary insights quickly. To give accurate predictions, Random Forest considers results from multiple decision trees merged together. Medical care has been improved through application of this approach, which predicts if a patient is likely to have a particular ailment using medical history and symptom evaluation. Use of said algorithm has increased accuracy in diagnosis, resulting in better treatment options being suggested by physicians.

Additionally, K-Nearest Neighbors (KNN) is another style of machine learning conveniently entailing classification depending on association between data points. A real life application could entail determining reading preferences for customers at an internet bookstore by analysis similarities between past purchaser's interests. If two people enjoy similar books, chances are one will appreciate reading books preferred by another reader. This aspect makes KNN exceedingly useful and takes common, mundane, tedious facets that are solvable expediently.

With their overwhelming usefulness, machine learning algorithms are necessary staple routine entities. From forecasting home prices, to providing relevant research materials, these algorithms shift the paradigm of technology and its interactivity with human beings. By understanding how these machine learning algorithms operate, we can make informed choices based on reliable information.

DEEP LEARNING 2

The subfield of machine learning known as Deep Learning supersedes all other techniques today when it comes to learning from data. Its breakthrough advancements in tasks such as speech and object recognition, natural language processing and gaming cannot be overstated. The concept of artificial neural networks that replicate human thought inside machines was established during early years of AI research by scientists like Warren McCulloch who collaborated with a mathematician Walter Pitts on modeling logic operations performed by neurons.

Despite this foundational work, progress was thwarted by challenges posed by hardware limitations and unavailability of large labeled datasets until the turn of the century. Great strides have been made in deep learning algorithms with advancement in computing power that we can trace back to developments such as Backpropagation algorithm, which facilitated training advanced Neural Network architectures known commonly today through their usage namely Deep Neural Networks(DNN).

DNNs have since become an integral tool for real world solutions ranging from picture categorization, to translating languages or predicting geopolitical events among other uses. In this chapter, we shall stride through fundamental concepts encompassing deep learning like neural networks architecture along with Backpropagation algorithm plus its applications in fields such as finance, healthcare or the entertainment scene. Therefore, this serves perfectly both curious enthusiasts, gifted scholars or professionals seeking knowledge about AI developments. Deep learning has seen superfluous advancement in recent times and has found its way into various applications including speech recognition, image classification, and natural language processing.

One major advantage of deep learning is that it can learn from raw data without the need for manual feature engineering. What this means is that the model can automatically learn relevant features from the data which can be used for prediction. In addition, deep learning algorithms have high scalability making them capable of handling complex and large datasets.

Nonetheless, training models based on the methodology of deep learning can still be a time consuming as well as complicated task. This is because deep learning models require a large amount of data and computational resources to achieve accuracy. Additionally, optimizing the training process can also prove difficult as it may require specialized knowledge and expertise. Despite these challenges, deep learning has been widely embraced in various fields including healthcare where it's used in medical imaging analysis as well as disease diagnosis; finance where it's used for fraud detection and risk assessment; and robotics where it's applied in object recognition and control systems. In addition, trained models have emerged as one of the most significant developments in deep learning. These are already trained on massive datasets and can easily be tuned up for specific tasks by developers who leverage their knowledge to create new applications.

Finally, transfer learning is an exciting aspect of research in deep learning. It involves using knowledge learned from one task to boost performance on another task. Transfer Learning shows great potential for improving efficiency and accuracy especially in situations where there isn't enough training data available.

All in all, there's no denying that while developing such models presents some challenges, continuing progress will undoubtedly yield more exciting breakthroughs in future applications.

When individuals talk over each other during conversations, it signals a dearth of regard for the speakers' perspectives. This conduct displays self-centeredness, where one gives little attention and importance to others' opinions involved in the discussion. Practicing mindfulness in communication would involve fostering attentiveness toward every party within discourse. Here, all voices get recognized fairly without infringement from rivals' voices or insecurities about keeping up with perceived competence levels appended onto participants' remarks.

The complexity embedded within constructing deep learning models cannot be underestimated as it involves the creation of models capable of learning from vast amounts of structured or unstructured data generated daily in different industries worldwide. This text provides valuable insights on the process of building a deep learning model step by step. Selecting reliable data is an essential and foundational part of the process with data coming from different sources like videos, audio recordings, sensors among others. Essentially, the accuracy of any predicted result modeled by your selected algorithm will indeed depend on good quality data harvest.

After we source good quality data as a starting point, pre-processing is necessary to cleanse the obtained information, remove distortions initiated during harvesting and transform this information into a format that is workable for the machine learning models. The next phase involves designing your model architecture, which requires selecting appropriate types like Convolutional Neural Networks (CNNs), Recurrent Neural Networks (RNNs), and Deep Belief Networks (DBNs) for specific tasks amongst others. We then determine how many layers or hidden units would be important to include alongside activation functions, considering each deep learning model has its strengths and weaknesses.

Thirdly, training follows once you have designed an algorithm after acquiring good quality data using pre-labeled/re-labeled datasets for training stages whilst combining adjustments of biases and weights. This is done alongside repetition upon repetitive iterations until we achieve minimal error rates between actual outputs compared to predicted targets as our end goal.

Fourthly, testing and evaluation become relevant next in assessing constructed deep learning algorithms' performance when applied to completely new datasets that have not featured before. This helps us determine if regular updating or modification is necessary to improve accuracy levels while ensuring that overfitting does not occur; hence performance regularization measures can be initiated for optimal results.

Finally, we deploy our constructed algorithm into production by creating APIs integrating our newly structured algorithms onto software applications or production pipelines. This helps us to make predictions on entirely unknown datasets that require regular monitoring checkups thus easily guaranteeing system updates regularly, which are required to enhance already established accuracy levels. Developing a deep learning model is an intricate undertaking that necessitates an extensive understanding of fundamental concepts and technical tools used in the domain. Nevertheless, with adequate knowledge and expertise, one can develop models capable of resolving intricate challenges while delivering accurate predictions. Building a deep learning architecture requires creativity accompanied with technical know-how and problem-solving abilities. The design procedure comprises multiple steps; none can be overlooked as each is pivotal in ensuring success with the ultimate model. To begin designing your deep learning architecture, you must accurately define the challenge at hand by identifying pertinent information such as data types you will be working with, target variables to predict/classify as well as other essential prerequisites involving problem statements. For instance, consider developing an image recognition system capable of distinguishing cats from dogs; this would require defining images that will serve for training/testing purposes among others. After finely pinpointing requirements for your project's problem statement, comes selecting the most appropriate neural network architecture best suited to address your challenge. Several techniques are available based on their distinctiveness in pattern recognition skills required for certain tasks—for instance, convolutional neural networks (CNNs) commonly used on image recognition tasks. Selecting an appropriate neural network architecture is followed by defining layers within your generated framework. Creating connectivity between layers while deciding how many cells each should house-nest such layers should allow optimal usage to exercise accuracy while managing resource allocation appropriately. It's imperative while arranging these components at each stage in modeling designs; one carefully organizes these different pieces by prioritizing details concerning their relationship, always striving for peak performance from their product. Finally comes setting up performance testing training generally conducted on the model, using a predefined dataset to help the model learn to recognize patterns in data, by adapting neuron weight distribution and interconnectivity. So, after a requisite duration of time and computing power is determined sufficient enough to facilitate active learning, this subsequently enables accurate prescient predictions. To ensure success with new, unseen data inputs, once trained separately from its original set of inputs, it's necessary for a model's performance during evaluation to meet expected levels before deployment occurs into real-world applications. Developing effective deep learning architecture requires iterating repeatedly through different network structures, such as layer configurations, along with selecting various relevant hyperparameters that promote optimum initial performance provisions. Each subsequent iteration is intended to achieve greater accuracy and better generalization.

With an immense potential for impacting how we process information across differing spectral distributions, deep learning forms what many regard as an extremely intriguing subfield within machine learning research. Here we crucially use artificial neural networks possessing multiple filaments meant for processing data analogous toward human comprehension techniques that maximally optimize signaling outcomes via signal propagation among such networks in recognizing predetermined pattern sets predicting outcomes intelligibly.

Backpropagation is an essential methodology within deep learning as it enables the neural networks to learn from errors made in its output calculations. The neural networks will update their predictions, per synaptic weight updates, minimizing prediction errors throughout all layers of network computations adjusting interconnection weighting. This allows the neural network to analyze more correctness in its subsequent prediction efforts. Backpropagation improves neural network predictions by adjusting the weights based on how much they contribute to the error. Calculating this contribution gradient involves using the chain rule of calculus backward through the network from output gradient to input gradients for each weight change direction. Though versatile across fields like self-driving cars or vision processing and speech understanding, backpropagation requires substantial data quantity and computational resources. Specifically about computational intensity: larger networks with millions of parameters extend gradient computation time through operations like matrix multiplication that often necessitate billions or millions of examples needed for optimal performance. In addition to being challenging computationally, Backpropagation remains one of the most dominant methods used for training neural networks despite its computational complexity that presents significant obstacles at times during implementation. At this point though, plenty of software frameworks and libraries have surfaced such as TensorFlow or PyTorch besides others. This facilitates offering developers intellectual refinement amid developing & training Neural Networks while harnessing their full potential by tapping into Hardware Acceleration which significantly hastens computation processes.

Furthermore, this advantageous method has played a pivotal role throughout the history of machine learning being responsible for many break-throughs made over the past few decades. As technology progresses forward bringing about improvements including algorithms and AI capabilities, techniques like backpropagation, among others, will likely continue being the dominant force behind advancing machine learning research well into the future.

The concept of learning rates is essential in deep learning where they play an integral role in adjusting weights during back-propagation promptly. Failure to comprehend what this involves can cause adverse outcomes when dealing with neural networks' performance.

Just like driving on winding roads where we must carefully regulate our speed so as not to overshoot curves or miss out on destinations if going too slow, proper setting of these rates will help achieve impeccable performance levels for your network.

In simpler terms, hiking hills come into play by taking small steps instead of larger ones that could mean taking more extended periods while arriving at the pinnacle. Still, the final outcome is worth this as small steps allow us to navigate obstacles with much ease. On the other hand, larger steps may appear quicker but may result too often in missing crucial details and going off course.

Due diligence is required when tuning and experimenting with varying techniques aimed at guiding deep learning's optimization process. Some methods include using a schedule for such rates that decrease over time as networks approach convergence or adjust rates dynamically based on loss function gradients using adaptive learning rate algorithms like Adam or AdaGrad. In deep learning, an obstacle that experts often deal with involves determining an optimal learning rate which can vary depending on aspects such as the neural network architecture or problem being solved. Achieving this can be tedious since there isn't a one-size-fits-all approach, requiring iteration via hyperparameter tuning. Picking out an algorithm for training poses another challenge as various algorithms have distinct properties accompanied by varying requirements concerning a favorable learning rate combination—making testing essential for finding optimal combinations. An additional hurdle pertains to finding a balance regarding just how high or low to set the ideal learning rate—a too-high value may risk overshooting whereas too low hindering progress by entailing local minima or saddle points' hostility that obstructs reaching better solutions. Implementing techniques such as adaptive algorithms or scheduling comes in handy for reducing these sorts of barriers. When neural networks process incoming data, they use an activation function that either activates or blocks specific neurons based on their features.

Despite this critical step for accurate processing of information via neurons in a neural network, default settings for these networks do not include normalized input values between layers (which can result in poor performance during training). Input values that are significantly larger than or smaller than the expected range for neural processing might cause unstable gradients when comparing weight adjustments and loss functions—hindering efficient optimization. Batch normalization addresses these problems by normalizing inputs within individual layers through batch processing techniques. These techniques capture layer-wise means as well as standard deviations from already processed batches of information, thus stabilizing outputs and reducing inefficiency issues while optimizing overall performance. For those working with deep learning models, batch normalization offers a crucial technique to stabilize optimization processes by normalizing input layers in neural networks. This not only facilitates more predictable gradients for faster convergence during training but also higher quality results on validation and test datasets while preventing overfitting. Here, the models become too complex for valid predictions on unseen datasets due to fitting too closely with their training datasets.

By removing noises impacting outlier outcomes in respective input layers during training sessions of convolutional neural nets (CNNs), recurrent neural nets (RNNs), feedforward NNs paired with techniques like dropout or weight decay, batch normalization consistently enhances models' performance reliability.

Of noteworthy reputation since its introduction by Sergey Ioffe and Christian Szegedy back in 2015, researchers regard batch normalization as essential when managing problems likely encountered through deep learning strategies. These problems include difficulty handling exploding or vanishing gradients within very large NN layers, which accounted for highly unstable and very slow training before the development of batch normalization. Training deep neural networks became much faster and efficient with a breakthrough innovation called batch normalization, which stabilized gradients providing predictability for researchers working on complex tasks like image classification, object detection or speech recognition resulting in dramatically improved performance amongst those objectives. Batch normalization influenced many researchers to fund further research into improving network architecture reliability leading to notable advances within the realm of deep learning. Today, it has furthermore sustained its prominence as it's seen as standardized best practice among all types of neural networks.

A type of machine learning algorithm called neural networks has contributed significantly across numerous applications covering areas ranging from image recognition to natural language processing. The most integral part of this framework is activating a specific node through its activation function as it adds vital non-linearity benefiting enhanced acquisition abilities for the detection of complex linkages amidst input/output recordings.

In actuality, many real-life outcomes are non-linear, given their characteristics that define complex relationships between the factors that influence them. It is here, that activation functions prove advantageous by allowing Neural Networks to recognize complex neural patterns than those of linear arrangements surrounding numbers.

In a Neural Network, specific artificial nodes receive signals; thus in response produce output signals sent to adjacent neurons at the next layer, constituting nonlinear functions of input signals. These activities promote an enhanced capacity to recognize more intricate data pattern formations across diverse types of input data stimuli.

Deep learning networks utilize various types of activation functions with Sigmoid, tanh, Rectified Linear Unit (or ReLU for short), and Softmax serving as different function types exhibiting unique characteristics with specific diverse strengths discerned against typical inputs.

Activation functions play a pivotal role in deep learning algorithms since they foster accuracy levels enhancing the probability of accurate predictions or classifications attained using prior dataset acquisition methodologies.

Although artificial neural networks draw inspiration from biological neurons' structure and function, it is important to note that they are inferior concerning complexity and behaviors relative to their biological counterparts. However, similarities do exist between the techniques used by both entities' artificial activation methods introduced into Neural Networks by transforming input signal using techniques such as sigmoid etc., closely mirroring similar firing patterns found among assorted geospatially widespread real neuron populations. In deep learning, artificial activation functions are instrumental in capturing complex relationships inherent in data. While there are disparities between them and real neuron behavior, they commonly perform similarly where ReLU has linear responses up to a given threshold, while sigmoid produces gradual firing rates as input signal rises.

Optimization algorithms play an essential role by adjusting network weights during training. Stochastic Gradient Descent (SGD) uses loss function gradients to update weights while Adam leverages adaptive learning rates for adjustment under variation.

To prevent overfitting detrimental to model generalization capabilities dropout—where nodes drop out randomly during training—L1 & L2 regularization penalize loss functions that encourage simple models.

Deep learning encompasses several crucial concepts besides those mentioned above, critical to neural networks' design and functioning and will help you develop end-to-end solutions for diverse problems such as advanced issues like natural language processing, text or image understanding, and many more.

Deep learning technology has improved across many industries since its inception during the latter half of the last century; it is especially seen within healthcare—displaying exceptional accuracy working with complex data like images or sound analysis.

Neural networks are infamous for their demystifying properties through utilizing deep learning; although created back throughout the late 1940s and 1950s era, not until recent technologies did they see significant development. Through better accessibility and larger training dataset sizes, current advances utilize this technology significantly. One excellent example: convoluted neural network design for practical image identification tasks display a high degree of accuracy when recognizing objects on images. Such achievement is exemplified by Google's DeepMind teams, whose developed model correctly identified over 1 million sampled images with an accuracy rate above 96%.

Natural language processing (NLP) is another field where deep learning has found its footing through recurrent neural networks or long short term memory networks generating realistic text and speech samples. OpenAI's GPT 3 passed the Turing test excellently with award winning results showcasing the machines' ability to generate human-like text. Incorporating deep learning methods into healthcare has shown a great return; specifically within medical image analysis—assisting doctors in accurate diagnoses, maintaining ideal patient outcomes and predicting outcomes for certain conditions. However gathering large volumes of high quality data sets for readily accessible training poses a critical challenge for advanced systems like deep learning. Presently, researchers are developing innovative transfer learning methods using pre-trained networks that enable quicker machine learning even with smaller datasets. Deep learning has increasingly become an essential tool for solving challenging problems across diverse fields like healthcare, natural language processing and image recognition. Its major advantage lies in its ability to learn complex patterns automatically from data without human intervention or explicit programming.

To improve model performance, researchers have identified data augmentation techniques like rotation, scaling or cropping that artificially expand the dataset(s) leading to better accuracy and robustness. Notwithstanding these improvements, one primary challenge with deep learning models is their black-box nature which makes it difficult for humans to understand how they arrived at specific conclusions as regard different tasks

Researchers have developed various techniques aimed at improving the interpretability of these models such as visualization of activation maps, feature importance analysis amongst others. With remarkable gains made in areas like image recognition and speech synthesis, key applications simplifying activities such as self-driving cars; notable implementations include AlexNet trained at the ILSVRC 2012. Whilst challenges exist alongside efforts aimed beyond traditional programming methods toward solving complex problems, we can benefit immensely from advances expected in the near future arising from research in this important field. During developments of neural network models in relation to image classification around AlexNet's creation period, solutions typically fell under two types: shallow hand-engineered feature neural networks or deep-learning networks featuring pre-learned structure learned from unprocessed image data itself. AlexNet stands out among these innovations in that it merged both approaches. It pioneered highly effective deep learning procedures based on raw datasets alone while incorporating novel components like ReLU to improve training speed and overcome challenges presented by gradients needing substantial modification during such procedures. A massive dataset comprising 1.2 million images tested this system to achieve measurable results that exceeded high-end products available at the time. This confirmed that CNN applications within computer vision tasks not only held potential but also marked an impressive breakthrough at the time. This stimulated further study and exploration of automated learning systems based heavily on neural networks—not only within the field of computer vision, but in areas like medicine. Deep learning's prowess in medicine has helped generate more accurate diagnoses and earlier treatment options courtesy of algorithms developed to assist with medical image analysis. The application of deep learning techniques has yielded promising results in developing accurate predictive models for patient outcomes and flagging high-risk patients. Nonetheless, there are obstacles that need addressing to fully unlock its potential benefits. A notable challenge lies in obtaining extensive datasets with high-quality data inputs. This can be daunting particularly for startups or smaller institutions without access or resources to these requisite amounts of structured organized data assets as a foundation for training algorithms effectively through machine-learning techniques. In addition to this limitation are computational expenses involved in both initialization and running processes that can restrict its use within certain domains of medical research and development applications. Furthermore, the opaque nature of deep learning algorithms means it's quite challenging interpreting the decisions arrived at; once deployed, this poses thoughts around possible biases or ethical dilemmas around how such conclusive evidence could/should influence our decision making. The overall outlook on deep learning is positive considering its immense potential in solving intricate problems. However, vital gaps remain to be filled as the technique evolves along ensuring innovative application by leading researchers fueled by progressive investment from industries interested in multivariate data-driven methodologies.

Deep learning is one of the most fascinating branches in machine learning, which offers an unbeatable solution to complex issues ranging from image recognition to linguistic analysis by training neural networks that improve predictions with time. A neural network is like the human brain's interconnected neuron meshwork, where each neuron processes relevant information before combining to make sense of referenced data from the dataset corpus fed during training. The subcategory lists under deep learning include Convolutional Neural Networks(CNN), Recurrent Neural Networks(RNN), Generative Adversarial Network(GANs), Autoencoders and Reinforcement Learning(RL). CNN is ideal for Image Processing and Pattern Recognition projects, while RNN tackles appropriately on tasks that involve sequential pattern detection e.g. speech translation into another language. GAN follows an unsupervised algorithm that includes creating fresh images as well as detecting authentic generated images, Autoencoders compresses/eliminates noise while generating new data, whereas RL takes lessons from making a reasonable decision under positive reward tactics. The training procedure utilized here helps agents decipher optimal decision-making techniques that leverage rewards for better environmental adaptability. This particular technology finds its applicability in a broad spectrum of areas such as gaming platforms, robotics regimes or autonomous vehicle programming systems.

In light of the numerous strengths and shortcomings associated with every subcategory present within deep learning theory, we must choose an appropriate model very carefully with regard to specific implementations. With ongoing advancements in research and development efforts directed toward deep learning models— novel methodologies coupled with newly emerging subcategories will likely beget a plethora of discernibly inventive use-cases.

COMPUTER **3** VISION

Computer vision is a fascinating subfield of artificial intelligence that aims at enabling computers to comprehend visual data derived from their surroundings. Its use cases span across various industries: for example, facial recognition or satellite imagery analysis are among its practical applications.

The study itself can be attributed back to the early days of computing when originating algorithms were developed aimed at detecting primitive shapes and patterns present within an image. However, it was during the late 1980s/early 1990s that we achieved progress resulting from software advancement and computational capability allowing improved machine perception achievement. A significant milestone took place with ImageNet's introduction: this boundary-breaking annotated-image dataset supports training various computer vision models boosting groundbreaking accretion discoveries such as object recognition or image classification. By using deep neural network architecture with image classification tasks, a significant advancement was achieved resulting in widespread acceptance; making deep learning a fundamental aspect when it comes to computer vision research.

Breakthroughs have emerged within particular industries like natural language processing, facial recognition to object detection, thanks largely due to this technology. Success has meant that computer visions' scope has expanded to include self-driving cars, surveillance systems, and medical imaging among others.

Additionally, it offered significant potential beneficial transformations across other sectors including those within healthcare, manufacturing and retail. Now enabled with analyzing large amounts of visual data, computers can process it at lightning speed becoming stronger tools toward understanding interactions happening around us.

Practical applications following computer vision are numerous across multiple sectors—Healthcare, Retail and Manufacturing, for instance.

In healthcare in particular, disease detection is more straightforward and precise, thanks to advanced techniques made possible through computational analysis stemming from changes created via machine learning algorithms. Through automated medical diagnostics which utilizes imagery, many complex diseases that usually require diagnosis from highly specialized physicians are recognized sooner allowing effective intervention at an early stage while reducing healthcare costs. While employing technologies such as X rays, CT scans, MRI's, and ultrasound, vast amounts of digital are available for processing via computer vision algorithms. In this manner tumors, bone fractures, or other abnormalities in the pictograms may be examined accurately through modality conveyed onscreen which allows for complex disease diagnoses such as Alzheimer's disease or cancer to be made with an even higher degree of accuracy.

Finally, robotic surgery assisted by real time feedback via high resolution cameras coupled with computer vision technologies are capable of detecting key anatomical features during procedures enabling surgeons to perform minimally invasive surgical procedures. Thanks to advancements in modern technology like computer vision, personalized treatments are now possible based on gathered individual patient data. Algorithms utilized by this innovative technique analyze information such as medical history alongside genetic traits and lifestyle patterns to predict future disease risks accurately. Additionally, identifying appropriate treatments through extensive analysis ensures that medical intervention has meaningful effects on patients' health. This leads to improved care outcomes while significantly reducing overall healthcare expenditures due to efficient results delivered timely from early diagnosis facilitated via leveraging its unique features.

In recent years, computer vision has transformed the way retailers operate by helping them obtain valuable insights into consumer behaviors and market trends. These insights aid decision-making in optimizing operations management processes that lead organizations toward achieving business goals such as increased sales revenue.

The advanced technological capability of Visual Search enables automated product recognition while matching similar items available in inventoried stock, creating convenience for shoppers who can easily navigate through stores' catalogs with spontaneous ease whilst providing an engaging personalized experience best suited according to their preference.

Similarly, Efficient Manufacturing requires state-of-the-art quality control systems that keep production errors at bay. Incorporating Computer Vision-based analytical approaches ensures accurate detection of defects during production processes. This reduces turnaround times significantly as the organization gets prompt feedback about where they went wrong increasing operational effectiveness.

In conclusion, with cutting-edge computer vision technology that continuously advances itself, businesses can expect an even more enhanced future of greater efficiency with improved customer experience tailored by deep insights into the market trends, consumer behaviors and optimized operations. Manufacturing today faces increasing complexity and costs that threaten the bottom line. Fortunately, computer vision is entering this arena as a game changing technology capable of streamlining operations across industries. This tech provides new insights into product creation from analyzing video footage from production lines—allowing for improvements in almost every aspect of operations such as reducing waste and improving product quality while diminishing costs.

Additionally, computer vision plays an essential role in predictive maintenance by analyzing visual data from production equipment—allowing manufacturers early intervention potential to avoid excessive downtime due for repair related reasons further saving costs (and time). Overall, with the advances in computer vision technology, shaping the next generation of manufacturing will transform industries into driving more efficient and sustainable practices making it not just better for business but also better by doing right by our planet.

Autonomous vehicles require sophisticated mapping systems coupled with refined understanding of their surroundings in order to operate competently on public roads without accidents occurring. Thus arises a vital application i.e., computer vision—facilitating efficient navigation by utilizing visual data collected via cameras and lidar sensors while creating comprehensive maps updated carefully over time. This technology has proven indispensable for ensuring safe driving practices while advancing toward technological advancements required for fully autonomous driving systems, which can be relied upon with confidence by users.

Although computer vision has improved significantly in recent years, there are still several **challenges and limitations** that must be addressed.

One of the primary challenges is the difficulty of accurately recognizing objects and patterns in images through algorithm development. This challenge becomes even more problematic when dealing with blurred, poorly lit, or noisy images. Recent incidents highlight the importance of ethical use of computer vision and CNN technologies. For instance, Google Photos' flawed image recognition algorithm erroneously labeled a picture of two black individuals as "gorillas."

Another example includes Amazon's AI recruiting tool that discriminated against female applicants by downgrading resumes containing words commonly used by women. These situations underscore the critical need for diversity and inclusivity in computer vision datasets to avoid algorithmic biases that can lead to discrimination or prejudice. Although these technologies have immense potential for revolutionizing many fields, it is crucial to utilize them ethically.

Another significant challenge is acquiring large amounts of labeled data to train computer vision models such as those used in ImageNet. While ImageNet has been pivotal in developing computer vision algorithms, there is still a requirement for more diverse datasets capturing different objects and environments found in the real world.

Lastly, deep fakes represent another challenge wherein synthetic media using machine learning algorithms create convincing but false videos or image content. "Deepface" brings together "deep learning," (a type of machine learning), and "fake" into a technological phenomenon capable of manipulating digital content such as images, videos or even audio recordings. Using Neural Networks—a machine learning algorithm—a set of image or video patterns can be learned. Then, this knowledge can be applied to produce new content with similar patterns through "generative modeling," frequently creating realistic facial features and objects.

To create DeepFake videos, the algorithm scrutinizes movements, and expressions in a dataset of the person(s) whose face will be used in the deep fake. Then it re-enacts these motions in another video that makes it appear as if such a person were doing that motion. In fact, they never did originally. On the other hand, for deep-fake image creation, the algorithm analyzes facial traits from a vast dataset of images featuring solely the person whose face will be included in new imagery to reproduce them under novel situations. This is done so that it looks like they're doing something which they might have not truthfully done.

Although these technologies could offer harmless entertainment options such as creating funny moments, there are concerns regarding their malicious potential use such as spreading false information or damaging an individual's reputation using fake content. Furthermore, there exist serious ethical concerns related to privacy violations from computer vision technology—the Facial recognition tech faces criticism for being invasive, leading to perpetuating bias and discrimination against susceptible individuals.

Finally yet importantly, although computer vision-based approaches pose certain criticisms, nevertheless their unique advantages cannot also just be ignored across healthcare sectors and also within manufacturing sectors—as it bears revolutionarily new Artificial Intelligence-driven potentials lately.

In recent years, the exciting field of Computer Vision has greatly drawn significant attention. Fundamentally, it involves designing machine interfaces capable of deriving insights-from images and videos taken out of their immediate environment-thereby identifying patterns or extracting valuable information required. This approach can help transform processes spanning varied industries like autonomous vehicles, facial recognition, onwards—thus it is proving to be increasingly versatile with each passing day.

Without doubt, Computer Vision is a crucial resource across different industry domains. For instance, in healthcare, it aids medical imaging analysis. In agriculture, it facilitates crop health monitoring and defect identification while the manufacturing industry utilizes it for quality control practices. Retail applications are focused on inventory management plus creating customer analysis reports. In entertainment realms, it is employed to render virtual and augmented reality viewpoints whereas security personnel utilize it primarily for identification procedures while continuously monitoring an area.

Tracing the roots of Computer Vision takes us back to the 1960s when scholars began exploring means of using machines to process visual information. The early focus then was on digitizing images and detecting simple shapes. But the central growth came much later, during the 1990s, primarily due to robust hardware and software advancements at that period, which accelerated advanced algorithm-driven image processing. Thus, paving ways for rapid innovations seen today.

Among noteworthy trends is deep learning where predictive computing occurs due to massive volumes of data being channeled into learning processes. There have been breakthroughs recently viable for object recognition, detection plus intricate image segmentation tasks—with Face or Shape Recognition after significant advances leading to current progress are computed.

Nowadays, the present frontiers expand further: Think self-driving cars enhanced by computer vision applications amid medical imaging operations—the future looks promising with continued innovation in this regard. Researchers are focused on developing algorithms that operate at an efficient level when analyzing images or videos while simultaneously discovering new avenues where advancements in digital picture recognition are well applied like virtual reality programs along with the field of robotics among others. The hurdles faced during this journey range from problems surrounding data quality to necessary improvements regarding explainable Artificial Intelligence (AI). Still, despite all this, certain developments provide hope for the future with regards to digital image understanding done right.

As technological advancement accelerates further, we can expect even higher levels of abilities providing us with advanced options across industries, which at present might not be available, yet whose impact will bring forth elevated efficiency levels within present industrial parameters characterized by safety and functionality.

Robotics equipped with the latest computer vision techniques have several benefits. The visual system allows them to perceive their environment, react suitably in seemingly hazardous conditions such as the factory assembly line or in unknown areas providing necessary assistance. The healthcare system has also included this technique for various purposes like disease diagnosis, medical imaging and drug discovery leading to an accurate diagnosis with the hope for a speedy recovery.

The entertainment industry also sees a significant impact of computer vision techniques, particularly regarding gaming experiences with more natural interaction of the user supported by realistic 3D environments and characters followed by necessary movements orchestrated by motion detection cameras that seem more lifelike than ever before.

History of Computer Vision

In the Sixties, came an exploration of computers interpreting visual data leading to one's ability toward Computer Visionary skills development. This was before today's more sophisticated capabilities with various types of pattern recognition within an image processing system or program evolved over time through dedicated tech. Object recognition relied on edge detection or basic geometrical shapes during earlier times in this profession until breakthrough advancements such as CNN in multidimensional analysis were exposed later through deep learning discoveries. This further increased adaptation abilities for unreal comparative accuracy between human-PC systems. Present research delves deeper into AI while creating enhanced learning techniques from larger sets upon development of better focused digital ecosystems supporting variety within Computer Vision environments, that primarily focus on multiple sensors, with AI being able to detect humans independently as well navigating physical environment surrounding it. This helps with movements correlated toward better understanding individual consumers' interests whether residential entertainment centers or even healthcare guidance processes backed by growing software tips and tricks shared on home computer message boards covering the next possibilities of computer technology. It's important to note that rejecting the truth about climate change goes against what science tells us. Existing empirical evidence derived from different sources clearly shows that our planet has experienced an unbroken warming pattern during the past century or so. Historical temperature data, along with other dependable metrics like ice cores or tree rings all point toward a unanimously accepted notion where humans are responsible for rising temperatures via practices like deforestation or energy production through burning fossil fuels.

Image Processing

New automated processes like animation rendering require powerful object recognition/ tracking tools to produce high quality even under real time scrutiny quickly. For example, Viola Jones' AdaBoost based.

The image scanning technique deploys multiple classifiers to scan images with different window sizes until locating specific features. These illuminating features are combined using a weak/ classifier approach producing a strong decision process yielding low false negative rates. By contrast, YOLO is an easily enhanced deep learning framework that subdivides graphic representations of real-world objects into grids. Here, each component is safe and can be quickly processed by neural networks with high accuracy because of fewer processing steps.

Despite the numerous algorithmic differences between Viola Jones and YOLO detection/recognition techniques, all the frameworks share a common feature extraction, image recognition and categorization module for discovering elements like edges, corners, among other salient object attributes. The solution to object detection lies in various algorithms such as Faster R-CNN—one technology developed by Ross Girshick's team back in 2015, hinged on deep learning alongside Region Proposal Networks (RPN). Its approach generates a proposal for regions to pinpoint any elements present within static images that later leads into detecting these elements accurately through classification via a Region-based Convolutional Neural Network (R-CNN).

Faster R-CNN enjoys rapid recognition within computer vision due to its precision grounded on real-time performance while detecting other objects during processing. Therefore, modern computer system architects integrate it into their systems for image recognition and tracking in diverse business cases.

With increasing advancements in image processing techniques, we get to unveil hidden information within images and videos beyond ordinary human interpretation. It includes preprocessing techniques such as Filtering, edge detection, and morphological operations that improve an image's quality before further analysis or feature extraction methods like SIFT, SURF, HOG to extract descriptive data features.

Detection algorithms such as Faster R-CNNs is one of the advanced approaches employed currently in computer vision emphasizing learning from several images abundant within diverse systems. The paradigm predicts that more advanced software abstraction devices implemented through constant research will arise for much more intricate machine learning tasks and applications.

Introduction to Machine Learning Algorithms in Computer Vision

Modern computer vision is built on machine learning, which grants computers the ability to recognize patterns and make decisions using data. There are several machine learning algorithms utilized in computer vision, each with distinct strengths and weaknesses. In this section, we will explore some of the most widely used algorithms in computer vision along with their applications.

One such algorithm is Decision Trees, which work alongside Convolutional Neural Networks (CNNs) to classify images or carry out object detection tasks. A decision tree algorithm operates by repeatedly dividing the input data into smaller subsets based on different characteristics until a decision can be reached. Each subset is divided based on the feature that separates the most significant amount of data and reduces impurity or entropy best.

Decision trees combined with CNNs can analyze CNN output and use it for future predictions as a decision tree ensemble; this allows different decision trees to work together for more accurate predictions. For example, they could analyze features of objects in an image to refine an object detection task's result picked up by a CNN.

Support Vector Machines (SVMs) are another machine learning algorithm that can be utilized in computer vision tasks using CNNs. SVM is frequently used as a classifier for image classification and object detection by finding the optimal line or hyperplane that divides each class of data. In image classification, SVM can differentiate between various categories using specific image features obtained from them. In conclusion, combining decision trees with CNNs support analyzing interdependent data improving predictive accuracy and efficiency in computer vision tasks' commonly used algorithms like SVMs. CNN is one of the most widely used deep learning algorithms across a range of applications involving computer vision. This algorithm draws inspiration from how neurons in our brain's visual cortex work. The organization around layers recognize varying features in visual stimuli through convolutional filters and subsequent pooling layers for reducing dimensions before passing through fully connected classifiers or regressors. Recall that Recurrent Neural Networks employ a mechanism that allows them to utilize previous input data when anticipating future patterns. The secret is in their capacity to remember preceding inputs and derive insights from this internal storehouse. Since RNNs are well-equipped for managing input information with varying lengths, they are uniquely suited for image sequence processing tasks.

Generative Adversarial Networks (GANs)

Generative Adversarial Networks or GANs are a cutting edge deep learning model designed to create new original content like images, music, and text. In essence, GANs consist of two parts: the generator and the discriminator. The generator's role is to create content while the discriminator functions to distinguish between genuine and generated content. In generating content, the aim of the generator is to produce a piece that will appear authentic enough to fool the discriminator into deeming it real. To initiate training for this process, random noise is fed through from the generator to the discriminator. Generated material is subsequently classified as fake or real by the discriminator with feedback being sent back to improve on weaknesses within both models respectively.

Over time, this iterative process optimizes both models' parameters resulting in better generated outcomes that may often match genuine ones. Loss functions are used to assess disparities between synthetically generated media and authentic ones during training sessions updating parameters in a bid toward plausible synthesis output. With such capability being achieved, GANs act as powerful tools in creating novel artistic elements lacking any prior examples leading to innovation brought about by AI creativity. Nevertheless, caution must be taken when it comes to discerning ethical implications relating to the use of such technologies responsibly; especially since unchecked usage may lead to manipulative deep fakes creation exploiting online audiences for nefarious reasons. 2: StyleGAN:

StyleGAN refers specifically to an improved variation of GAN capable of generating images highly resembling human artwork that seamlessly span multiple styles or categories such as portraits of landscapes. This is achieved through an adaptable generative network paired with a unique class activation scheme thereby representing actual features crucial in composing high quality synthetic imagery. This approach has contributed significantly toward photorealism's progressiveness with its utilization today.

CycleGAN is a commonly used type of Generative Adversarial Network (GAN) that has a specific focus on converting images from one format to another. This is achieved through the use of two generator networks and two discriminator networks, each with their own responsibilities in the process. The generators are responsible for converting images from one domain to another while the discriminators ensure that the final product appears realistic and belongs to the desired domain. DALL E is another GAN based model that is able to create images based on textual descriptions. Instead of using multiple networks, this model utilizes a transformer network to convert text into a latent vector before passing it onto the generator network for image creation. BigGAN differs in its purpose as it primarily focuses on generating high quality images with significant levels of detail and resolution. This model makes use of a generator network with increased parameters enabling enhanced complexity and increased realism. Additionally, BigGAN uses loss functions that encourage it to produce realistic images fitting certain categories or styles. StarGAN offers multi domain image to image translation capabilities by using multiple discriminator networks alongside a single generator network. Each discriminator aims to ensure that produced images belong to specific domains while keeping the other domains out.

Finally, Deepfake is another GAN based model with a somewhat infamous reputation due to its ability to create fake videos, which convincingly depict individuals doing or saying things they never did in reality. Despite ethical concerns around this type of technology's potential use for creating fake news and misinformation, they also offer exciting creative possibilities in industries such as film production and special effects. The number of GAN variations keep growing as deep learning continues advancing. Given how quickly this field is progressing, it's safe to expect ever-more capable GANs serving multiple purposes.

Deep learning makes use of transfer learning and fine-tuning to extract value from pre-existing models during each assignment's workflow setup process accordingly.

Computer vision workload requirements have led many teams toward these two methods since such expansive data sets demand too many resources when starting with no available knowledge base established beforehand. Instead of vying for intensive training investments required for the creation of new models entirely from scratch, experts utilize these learned features such as edges, corners, and curves found in more established models like VGG or ResNet.

Regarding fine-tuning specifically, pre-trained models can be augmented by adding new layers to their architecture. This improves performance on new struggles since the old layers stay unaltered and retain the existing learnings. Deep Learning technologies offer powerful strategies such as Transfer Learning (TL) and Fine tuning (FT), which significantly decrease time and resources taken up when starting development from scratch, while providing an additional set of features for labeling tasks down-the-line, such as medical imaging or conservation efforts. Implementations involve preserving initial layers present within more traditional pre-trained models but incorporating advancements learned within specific datasets through newly constructed fully-connected operations.

In terms of computer vision applications implementing TL/FT methods, Google's use-case within Medical Imaging identified successful diagnoses accurately via efficient identification of Breast Cancer incorporated into established mammograms. Waymo has invested FT in their self-driving assists by using similar methods to identify safety concerns within their autonomous driving technologies such as identifying bicyclists, pedestrians and other vehicles on the road network.

Overall, TL/FT techniques offer a highly scalable solution for generating value across sectors such as agriculture—helping farmers to detect crop-impacting diseases—and conservation efforts through advanced object tracking methodologies for protected species like Tigers. The implementation of TL/FT methods will continue to be a useful tool for addressing real-world problems where large datasets need querying or ongoing adaptation without the need for repetitive automation practices carried over from earlier training cycles.

Experts at the University of California conducted a study that involved applying transfer learning techniques to train a pre-trained deep neural network model for identifying diseases in tomato plants. As anticipated, the improved model exhibited striking accuracy when detecting diseases and could help farmers take swift action aimed at mitigating crop losses.

These examples demonstrate how applying transfer learning and fine-tuning concepts lead to enhanced efficiency and better accuracy for computer-driven surveillance systems across fields in diverse industries. Machine learning algorithms have been instrumental as solutions for intricate data classification issues faced while improving image generation accuracy from decision trees all through sophisticated deep neural networks.

The evolution of machine-learning powered computer security research only grows alongside advancements in industry-wide adoption circles continuing even now if we desire significantly better outcomes going further ahead.

Application of computer vision

Computing Vision has grown explosively popular recently as various industries have leveraged its versatile applications comprehensively including recognizing facts through visual data or enhancing autonomous developments required for self-driving cars. In this section, we will explore Computer Vision's expansive capabilities to find extensive information on different approaches such as object recognition/tracking; face detection/recognition; optical character recognition (OCR); medical imaging/analysis;

drones/autonomous vehicles; augmented and virtual reality (AR/VR); surveillance/security measures; robotics/automation processes, environmental monitoring and agriculture among others.

Object Recognition and Tracking represent crucial elements within computer vision technologies due to their ability to identify objects accurately and then track them within large datasets and video streams. This technology has numerous use cases such as securing properties more efficiently preventing accidents with autonomous cars than relying on humans using non-stop analysis processes or refining operation models of drone services efficiently.

Face Detection and Recognition are integral components that are related to leveraging computer voice functionalities performing vast operations such as understanding individuals' facial characteristics faster. It is useful for a range of applications across different industries, including security/surveillance measures or identifying individuals' biometric data accurately.

OCR provides advantages for quickly converting lines of printed text into digital materials that operate much efficiently in our current environments. This powerful technology eliminates human errors that arise out of reading countless pages conducted over monotonous and repetitive tasks related to journaling information through paper-based systems.

Medical Imaging Analysis adoption has taken root concerning computer vision technologies in recent years as medical institutions aim to implement highly accurate diagnostic capabilities seamlessly. Medical experts rely on algorithms based on processing large volumes of medical images such as MRIs, X-rays, and CT scans identifying cancerous tissues analyzing masses quickly. The versatility with which computer vision has been utilized continues to grow successfully across various sectors upholding robust progress through myriad applications such as autonomous vehicles and drones exemplifying their utility, which are significantly critical for industry transformation. It effectively detects obstacles, safely avoiding pedestrians, enhancing navigation systems by producing high-precision maps that is valuable autonomously. Augmented and virtual reality abound as further among several innovative areas upon which computer visions align, enabling optimal results from performing real-time object detection, heightening immersive experiences that are at once unparalleled.

Notably, critical examples apply surveillance operations and security initiatives making them efficient with unprecedented precision targeting suspicious activities more swiftly, without replicating undue costs in monitoring public spaces effortlessly. Robotics and automation become more efficient by relying on computer vision technology to perform tasks through object tracking with high accuracy, detecting product defects at the quality-control stage.

However, several challenges exist with data quality, shaping the foundation for accuracy as a significant obstacle necessitated through high-quality data searches struggling to administratively address biases, inaccuracies or inaccuracies that affect algorithm performance in real-time. Additionally, privacy concerns pose issues arising from multiple implications of computer vision technology's use. This further emphasizes secure information storage and management according to established ethics that manage privacy rights issues, while advancing crucial necessary technology. Despite increasing surveillance capacities at their disposal, it may impede ethical advancements traced to biases emerging in certain cases. Computer vision algorithms mainly rely on the quality of their training data. Thus biased sets may give way to unjust outcomes like discriminatory predictions from facial recognition systems that are unable to identify individuals with darker skin tones due to insufficient authentic diversity present in their datasets.

Researchers continue exploring different options helpful in surmounting such challenges by devising new ways, improving existing techniques alongside producing more accurate and fair computer vision models. Two such approaches so far include adversarial training and explainable AI techniques, both serving as a tool to increase the robust nature of computer vision algorithms toward data variation and transparency.

Apart from these emerging trends, there are others shaping the future of computer vision. For instance, 3D reconstruction tool creation allows the modeling of 3D objects or scenes using two-dimensional images most relevant to applications in robotics. Here, robots' movement planning could involve detailed spatial information imparted by accurate 3D models created with this method. Similarly, video understanding tools for surveillance analysis and security purposes among other things provide an essential role by helping detect anomalies or identify potential safety threats using video data studying mechanisms.

Finally, there's the rapidly emerging trend of making AI systems more comprehensible through explainable AI techniques simplifying machine learning models to lay persons' comprehension levels; for example:

Convolutional Neural Networks (CNNs) remain relatively complex architectures despite efforts by researchers over time. The CNN has multiple layers consisting of various processing mechanisms like pooling or convolutional layers triggering feature activations within high-dimensional spaces complicating its ability to classify an image correctly. To further complicate matters, the tracking of individual weights and biasing parameters adjusted during CNN training with complicated procedures remain challenging despite researchers' best efforts. To comprehend the complexities of CNNs effectively, researchers have invented diverse approaches for interpreting their inner workings visually. One technique employs highlighting filters learned by the first convolutional layer that exhibit substantial tendencies toward specific features like edges or corners observed through visualization. Another method resorts to generating activation maps for intermediate layers that illustrate distinct detectable traits on each layer. Lastly, analysts offered gradient-based ways that help determine which image regions contribute significantly toward a particular class output neuron known as saliency mapping or class activation mapping. This method becomes a potent tool for gaining insights into network decisions when properly applied. These methods can help tackle challenging instances in model performance diagnosis prone during prediction procedures associated with challenging tasks related to analyzing medical images or monitoring crop growth seen across several industrial contexts. It's always essential that we consider new techniques that embrace new perspectives on procedures guaranteed at uncovering more robust applications while exploring ways of handling issues found in this realm positively.

NATURAL LANGUAGE PROCESSING

ntroduction to natural language processing (NLP): definition, history, applications, and challenges

Natural Language Processing (NLP) is a fascinating field of computer science that deals with the interaction between machines and human languages. Essentially, it's about teaching computers how to analyze, understand, and generate natural language text using algorithms. Thanks to advances in machine learning and deep learning in recent years, NLP has become incredibly popular. In this chapter, we'll explore NLP in detail—its definition, history, applications, and challenges.

Definition of NLP: At its core, Natural Language Processing means that computers can understand and work with human language just like we do. However, unlike humans, who innately have this ability due to their cognitive skills developed over time through experience and education, machines require programming through algorithms and models to accomplish tasks like language translation or sentiment analysis. These tasks involve analyzing emotions or opinions expressed within the text structure. As a result of NLP research on these algorithms and models through the years, long study of artificial intelligence has made it possible

45

for a machine or computer program to interact with users by generating natural sounding responses just like humans.

History of NLP:

The history of NLP can be traced back to the 1950s, when artificial intelligence was first emerging as an academic discipline. The early progress in this field concentrated on developing rule based systems, which could process natural language input consisting of words and sentences formed by people. Such systems were restricted due to their lack of ability for scoping out complexity and variability, while handling human languages effectively; thus proper execution needed more techniques added over time.

Joseph Weizenbaum's ELIZA program is one such example from the 1960s, which depicts how the results of rule based programs' limited ways only improved up till more sophisticated techniques were developed eventually for use, making Natural Language Processing simpler than ever today! Back in the 1980s, natural language processing (NLP) had a significant discovery when researchers developed statistical-language-processing methods. This particular approach enabled computer systems to detect patterns in language data so that predictions could be made based on said patterns. As time progressed into the 1990s, new concepts emerged around studying large text collections known as "corpus linguistics." This facilitated trend analysis and pattern identification incumbent to specific forms of language use cases that helped researchers build robust language models applicable across various NLP applications.

It became more apparent throughout the early 2000s that growth in internet ecosystems reflected larger scopes for natural-language-processing-based web-applications useful for diverse purposes like search engines, spam filters or social-media-analytics. The advent of "web crawling" together with "sentimental-analysis" techniques suited to extract meaningful insights from large-scale online content transformed this field with innovative approaches yielding outstanding results we still see today.

As technology continues advancing at an increasing rate, it allows machines to become more sophisticated and capable due partly to traditionally unmonitored massive amounts of textual data. Additionally, innovative techniques are continually being researched like deep learning algorithms together with neural networks, generating very positive results.

1. In several industries today, **NLP applications have become fundamental**.

    One primary illustration is customer service, where we have advancements enabled client-company interactions by leveraging powerful tools. NLP techniques are being leveraged across several industries to enhance customer experiences significantly. One such example is chatbots—AI-powered conversational interfaces that enable us to communicate with businesses using natural language—which have emerged as a popular tool among companies seeking improved consumer engagement through their websites or apps. From Bank Of America's banking assistant answering common queries, we may have on accounts/services online Shopping Assistance Chat Bots/NLU AI Apps For Big Brands(Example Domino's). All of these assist us with order/booking status. Voice Assistants like Amazon Alexa/Apple Siri/ Google Assistants (Nowadays found in most mainstream devices from phones/laptops/smart-speakers) leverage NLU plugins that offer personalized assistance merely via voice commands reducing physical interactions with consumers thus placing convenience on an all-new level. Additionally, NLP technology is utilized to analyze large areas of customer sentiment feedback, from polls and surveys to social media posts. The output of sentiment analysis provides valuable intelligence on customers' perceptions of a brand or product.

2. Across borders, NLP algorithms can be configured for instant language translation in real-time, eliminating language barriers in international business interactions. Automated email responses employing NLP have become increasingly popular with companies in addressing frequently asked questions. By using AI-powered algorithms, these systems quickly analyze the context of customer inquiries and requests and generate accurate responses that help resolve issues efficiently.

3. Natural Language Processing has significantly improved healthcare systems by allowing structured data extraction from clinical notes and providing readily available patients' medical records. EHRs are fundamental tools in achieving this, enabling medical professionals worldwide unlimited access to these critical records, as they identify patterns that improve quality care for every patient. The system enhances research outcomes with billing management enhanced as well.

4. Clinical decision support systems have also relied on NLP algorithms for personalized treatment recommendations based on patients' factors like symptoms or medical history. Application of this technology grants patients more control in managing their healthcare by utilizing chatbots powered off NLP-based algorithms for simple tasks like medication refills, scheduling appointments and attending best practices seminars on health, relevant specifically to their healthcare needs.

5. The finance industry heavily relies on the principles of natural language processing advancements. They are unmatched regarding obtaining insights from crucial sources of unstructured data such as news articles, social media posts amongst others, which help place bets

ahead into all sectors' markets outside ordinary financial metrics that we previously entered manually.

6. Through the same principles learned in fraud detection cases where anomalies were detected (phishing attempts), leading experts can now manage losses smoothly with early detection within insurance claim forms before approvals are granted This helps save significant funds from would-be fraudulent claims distributions that were placed at the expense of genuine customers in need of insurance assistance.

7. Customer service levels have also been boosted by institutions at large, thanks to NLP chatbots-powered responses to client inquiries and questions that help make better-informed investment decisions. The solutions together with personalized recommendation features provide a tailor-made experience for every client. This comes in handy as the world moves toward remote experiences for their clients, enabling automation results while managing its in-house operations tracking customer satisfaction.

8. Natural Language Processing has gone beyond the realm of sentiment analysis alone; its impact on trading strategies is both significant and powerfully leveraged in finance. Trading decisions have become smarter with algorithms parsing unstructured data that are gleaned from various sources such as news items and social media posts foretelling the likely impact of certain events on markets.

   Hedge funds are among the prime movers utilizing NLP insights: they use it to detect nuanced messages embedded within news reports before making either buying or selling decisions related to specific stocks.

9. Another marked application of NLP relates to governance measures that look into regulatory compliance activities such as risk management issues confronting banks.

Whether it is Anti Money Laundering (AML) or Know Your Customer (KYC) regulations, NLP processes are highly adept at analyzing financial reports, legal documents and news items thus providing us with warning signals of any potential missives.

10. The social media industry has benefited greatly from Natural Language Processing (NLP) technology, which has enabled platforms to better understand and analyze vast amounts of user generated content. One significant way that NLP impacts social media is through sentiment analysis, which allows algorithms to determine users' opinions of certain topics or products based on their social media activity. This information can be used by platforms to improve ad targeting and provide personalized content. For instance, Twitter's algorithm can identify sarcasm and irony to accurately gauge the sentiment of a tweet.

11. Another key application of NLP in social media is content moderation. The technology helps detect inappropriate or harmful content so that social media companies like Facebook can keep their platforms safe for all users. Chatbots powered by NLP have become commonplace on social media as well as understanding natural language queries and providing tailored recommendations and responses. Facebook's M assistant is an example that helps users plan events, make reservations and buy tickets.

12. NLP can also be harnessed to analyze images in videos posted on social media channels enabling descriptions for visually impaired individuals to be automatically added as it happens on Instagram. Trend analysis is another function it performs, which entails the identification of topics with great interest among users yielding improved advertising targeting via Twitter's NLP platform.

13. Additionally, this technology empowers translation into over 100 languages, making it more accessible on a global scale. The education industry also stands to benefit from the transformative potentiality of NLP in facilitating innovative teaching methods involving data comprehension. This is an educative development that is poised to usher in teacher empowerment as they effectively leverage these technological advancements.

Educational institutions are capable of providing personalized learning experiences to every student with the help of NLP algorithms. Intelligent tutoring systems analyze student performance data to identify where students are struggling, thereby creating personalized feedback and suggestions to improve their skills. A prime example is Carnegie Learning's MATHia platform that analyzes students' responses in math problems and provides personalized feedback. In addition, automated essay grading is a popular application of NLP that saves teachers a significant amount of time while giving students prompt feedback. edX's automated essay grading system serves as a good example for this.

Furthermore, NLP algorithms can even identify the learning preferences of the students by analyzing their performance data resulting in educational content that caters to individual needs. Knewton's adaptive learning platform uses NLP algorithms to analyze student data, grasping their learning styles and preferences accurately to create personalized learning experiences. Language learning tools like Duolingo language learning app also employ these algorithms to provide feedback on pronunciation and grammar.

Moreover, automated summarization is another area where NLP is being used extensively in transforming the education industry through concise summaries provided by analyzing educational content from various sources and identifying central concepts and topics from textbooks so that students can study more efficiently. The Highlights for High School project exemplifies such applications whereas platforms like Top Hat use NLP algorithms to provide interactive learning experiences from analyzed textbooks.

Lastly, question answering systems are being developed using NLP algorithms that analyze student questions and provide automated answers for swift understanding. The Watson Education Advisor from IBM is a noteworthy example of utilizing NLP algorithms for scrutinizing learner queries and offering answers.

Challenges in NLP

The field of NLP has made significant advancements over time. However, several challenges remain unaddressed even today which need our attention. **Some key obstacles encountered** while working **with Natural Language Processing** are listed below.

The primary stimulation here revolves around established quality controls for datasets used in training AI, where correctness, biasness, inaccuracy and incompleteness lurk around during development stages. Experts agree that the NLP algorithms require pertinent data for accurate functioning. Consider an Algorithm that is trained on the formal tone—it will not be able to interpret and process informal language in use across social media or casual conversations.

Likewise, if the dataset used to train an NLP algorithm exhibits signs of biasness, it might not represent the right groups of people causing discrepancies that limit accuracy and obstruct optimal performance. Essential measures such as curating datasets with care, selecting suitable sources that capture diverse real-life contexts and conducting thorough quality checks can alleviate data quality issues.

Effective handling of confidential information is a great concern in Natural Programming Language. Data privacy violations occur when sensitive personal information shared intentionally or accidentally gets exposed. Learning mechanisms mandate large personal datasets names, locations etc.; their careless handling can pose identifiable threats risking personal privacy.

Unintentional public sharing of valuable insights like political affiliations or health conditions on social media using NLP could lead to even more grave situations. The chances increase multiple-fold when unauthorized parties access it through mechanisms that are not secure enough. A solution here is implementing sturdy privacy guidelines and safeguards to sensitive data sources, giving impetus to secure mechanism usage. To protect individual's privacy while developing NLP algorithms, measures like data anonymization should be considered and implemented during collection of sensitive personal information used for training purposes.

Furthermore, proper storage and access controls must be in place ensuring the security against unauthorized personnel obtaining access to it while compliance with relevant regulations is equally paramount.

To mitigate potential biases arising from various sources such as inadequacy in dataset representation or algorithm design prioritizing specific types of language or sources of information, emphasis should be on diverse representation during training datasets selection. At the same time, we should ensure transcendency via transparent evaluations involving multi-stakeholder participation throughout its development life-cycle.

Natural language being largely open-ended, possessing temporal irony such as syntactic ambiguity creates confusing contexts that require in-depth understanding for proper interpretation. For example, "I saw her duck" bears different meanings based on contextual usage implying witnessing ducking action by a female or sighting a duck belonging to the same said person. Natural Language Processing (NLP) involves developing programs capable of accurately identifying word meanings using computer logic. This however, requires accounting for contextual variation—e.g., analyzing sentences within paragraphs within academic papers versus cheeky online memes. A critical challenge for NLP involves accurate interpretation due to several possibilities for word meanings depending on context. For instance, "Bank" could point toward financial institutions or physical features such as riverside edges.

Researchers identified several techniques such as named entity recognition, context analysis in addition to machine learning algorithms aiding Natural Language Processors with enhanced reading comprehension capabilities enabling them to narrow down what best fits specific cases being analyzed across most industries. These initiatives remain significant challenges for the researchers due to constant changes in industry standards or updates. Explainable NLP models are an intriguing topic for researchers today. These models aim to provide an explanation for their output, presenting users with the rationale behind their important decisions.

This essence is particularly vital in healthcare settings where it's imperative to understand how diagnosis or treatment recommendations were arrived at by machines. Additionally, cross-lingual and multilingual NLP has gained popularity over time, and seeks advancement models that can comprehend and coordinate languages passing beyond the realms of English. Potentially, it will not only improve accessibility to information by people who are non-speakers of English, but also facilitate efficiency when communicating across nations.

The extent that Natural Language Processing could change our way of interaction with differentiating language and information cannot be underestimated. However, the existing challenges include issues around data quality, bias explainability among others. As researchers persistently work on developing inventive techniques, major ear-to-the-ground advancements remain high.

Fundamentals of linguistics and language modeling: phonology, morphology, syntax, semantics, pragmatics, and discourse analysis

As key components of our daily existence, Natural Language Processing (NLP), which analyzes processes between computers and human language, has developed rapidly over time. To accommodate this intricate relationship, it's critical that we have extensive knowledge on linguistics and cohesive blueprint designs reliant on sound language modeling. Beverhey chapter focuses on fundamental concepts that deal with basic linguistic understanding by examining constituent elements including phonology, morphology, syntax, semantics, pragmatics and discourse analysis, which are essential for coherent communication.

On a greater scale, phonology is one core focus area in linguistics that concentrates on sounds within varying languages. This encompasses organizational patterns used within different contexts that form basic building blocks upon which multiple sound units contribute toward creating differentiation. The smallest unit is the "phoneme," serving as building blocks contributing toward different unique sentences/messages ultimately conveying meaning via inflections as well as expected cues.

Admittedly, studies apart from phonetic composition only enhance significance through other components like intonation or rhythm for added expressive effects with profound semantic cues beyond just context, especially when leveraging varying dialects amongst reference persons or cultural dimension groups globally. Language acquisition and learning would be incomplete without considering the crucial role that phonology plays. Children rapidly learn how to recognize and articulate their first spoken tongue's sounds from an early stage—this fundamental process underlies developing communication abilities. In addition, success in communicating effectively when learning another language entails mastering its specific phonological patterns.

The study of word structure forms part of morphology; it addresses how words come together and relate to each other within languages by exploring morphemes—smallest units carrying meanings—determinative within them. Morphemes combine into larger units like nouns or verbs from which the meaning itself arises; they can also signal variation in tense or genre as well as intonation changes in some cases.

Notably, NLP relies on sound knowledge of morphology given that it enables understanding complex relationships between different morphemes that make up meaning across natural languages. Syntax establishes regulations for forming phrases, clauses, and sentences by organizing words with intention regarding order and logic to form grammatically correct statements, whilst conveying meaning effectively. Therefore, adherence to standard syntax principles as per guidelines to respective languages ensures that communication between individuals remains properly understood. We can implement complex ideas or clauses into a singular sentence, which can be made possible through syntax with use of sentence connecting or conjunctions. In various linguistic phenomena like word meanings and sentence meanings, semantics plays an essential role in facilitating communication between individuals who share the same language. It encompasses the meaning of larger units of discourse constructed through context, inference and pragmatics.

The significance of semantics is evident in vocabulary acquisition where gaining insight into contextual usage helps learners apply them accurately while conversing. In computational linguistics too, researchers use natural language processing techniques to extract information about usage patterns within text datasets. It's made clear by now that speakers will, at times employ non-literal language like metaphors and irony in order to elevate the experience of their expression beyond its literal interpretation. However, non-literal expression can require listeners tactfully applying themselves in considering context and tapping into personal experience/knowledge, so as not to miss the intended message/purpose behind what was spoken. That said, our conventional spoken guidelines apply just as much here, especially around how we use contextualization when we express sensitivity toward an individual/audience faced with difficult topics/requests. Consequently, a level of social/moral etiquette is needed to achieve the right balance for newly introduced topics/content.

Adhering to Pragmatics (i.e.: being aware of these nuanced communication norms) allows speakers to communicate more effectively across different societal/ cultural contexts and to create connections that can bridge old gaps. For example, when learning and teaching language itself, it's important that learners appreciate the importance of pragmatics in contextual understanding and use words/phrases beyond just their respective meanings. Knowing the adequate way to deliver a request, or how you show gratitude in this culture vs. another will give students confidence when using language in social settings.

Further, study into communication practices especially within psychology has provided us with fascinating breakthroughs about how human interaction works on a larger scale. By teasing out all the formality behind individual dependent contextual conversation practice, we can start capturing various cultural expectations, and discover fundamental perceptions about society's underlying behavior through commonly used contextual gravitations.

The study related to Discourse Analysis focuses on analyzing contextualized usage through writing or orally generated content used primarily across various domains like politics, media, life experiences, within societies and groups. The primary goal here is that we interpret where emotions or appeals may have been employed to convey meaning or a particular stance.

The analysis of language use can help us establish and understand different links woven between context, identity and power. Linguistic patterns often get shaped by our cultural norms, historical events, political context and social factors at play, which get reflected through language in interactions.

Discourse Analysis also consists of various aspects such as how content/theme is discussed or agreement/disagreement conveyed—it's sequencing. One also has to look into the specific linguistic tools employed to signify a certain hint.

There are several applications associated with discourse analysis; typically used for subjects like sociology, anthropology, media studies and education where one can explore cultural significance.

It helps us understand societal challenges by analyzing news coverage and presents the impact that language holds over society and politics.

Language is mobilized to create meaning that aligns with shifting social structures, which gives rise to diverse identities and relationships, providing broader avenues for research in this area.

Teaching second languages using discourse analysis can provide benefits as learners would be exposed to genuine usage enabling them in comprehending the maneuvering of Social Perspectives while developing appropriate usage skills across different forms of interaction.

It's prudent for anyone aiming at building NLP systems facilitating accuracy when it comes to Natural Language interpretation/generation to have fundamental knowledge relating with phonology, morphology, syntax, semantics, pragmatics and discourse analysis. Such areas offer assistance in establishing an initial base within NLP technology development. The current administration's measures have proven contentious among citizens within our nation. With various sections of society expressing divergent perspectives, some contest that these policies violate core American principles whilst others praise them for dexterously advancing interests for positive change.

Preprocessing techniques for NLP: tokenization, normalization, stemming, lemmatization, part-of-speech tagging, parsing, and named entity recognition.

Within natural language processing, proper preprocessing is paramount for turning raw data into machine readable formats that algorithms can readily analyze. The central theme in this section revolves around examining valuable preprocessing methods widely used in NLP applications such as text classification, sentiment analysis and machine translation among others.

Tokenization lies at the heart of numerous NLP processes such as splitting texts into smaller units called tokens represented by words/subwords/phrases/characters delimited by whitespace and punctuation marks after eliminating any unwanted characters like symbolics or punctuations—one good example could be email addresses. When tokenized, the email address breaks down to separate domains from usernames enabling accurate categorization. Once text has been tokenized, it becomes easier for machines to conduct tasks such as retrieving information or performing sentiment analysis since tokens can be uniquely identified and analyzed.

Normalization is an additional crucial aspect of NLP preprocessing as it ensures uniformity across diverse input data sets. Normalization takes several forms, including stemming, where specific parts of words or domains like emails, typically prefixes and suffixes are modified or removed to create simpler forms that computers can readily analyze, integrate with other datasets. Stemming refers to an essential technique used for reducing words down to their most simple form, commonly referred to as 'root' or 'base'. Using this method involves removing any suffixes from words, which leads them back toward this simplest version— for example, running becomes run while jumping reverts back into 'jump'.

This practice is essential due its effectiveness for lessening vocabulary size and boosting computational efficiency when applied on natural language processing tasks like sentiment analysis and text classification. Removal of suffixes from words constitutes a key objective of stemming algorithms widely used for natural language processing tasks like machine translation or sentiment analysis.

The Porter stemming algorithm and its successor Snowball employ different rules in achieving this end with Snowball covering several languages thereby making it more flexible than Porter. Notwithstanding this advantage however, there may be situations in which stemming outputs non-words thereby compromising task precision; e.g., applying Porter on "argument" generates the non-word "argu". To avoid such issues, one recourse available is lemmatization—a more sophisticated technique that analyzes word structure prior to carrying out reduction but still retains meaning: thus, say the lemma instead of stem for "better" yields "good", given similar sense conveyed; similarly "mice" reduces to singular form "mouse". Although Time consuming compared with Stemming, Lemmatization has proven effective in improving task Accuracy. One example is WordNet Database containing numerous terms with corresponding Lemmas while spaCy and NLTK employ sophisticated algorithms in detecting base forms. In natural language processing (NLP), part-of-speech (POS) tagging is an important step that assigns a grammatical category—noun, verb, adjective or adverb—to each word in a sentence. This helps us comprehend sentences better by identifying roles of individual words in expressing meaning or bringing out patterns in our text data. For instance, let's consider, "The cat sat on the mat." With POS tagging, we can tell that "the" functions as determiner while "cat" happens to be a noun; "sat" plays some verb role while "on", "the," and "mat" are identified as preposition/determiner/noun respectively. To achieve this task, statistical models were trained using large datasets of labeled text that utilize context of neighboring words, sentence structure etc., for predicting correct parts of speech for words. Afterward, applying tagged corpus proves beneficial in NLP tasks like machine translation, sentiment analysis, named entity recognition, classifying sentences into categories and extracting necessary information from large bodies.

When we analyze spoken discourse under Linguistics topics, Parsing evaluates their underlying structures for all relevant meanings. By doing this, Parsing isolates individual components for more straightforward analysis and correlation into cohesive grammar.

Parsing primarily subdivides language content into components under class designations. Here, grammatical structure is further projected through a tree-like buildup; tracing the relationships between each spoken word's related parts-of-speech elements within its organization.

For Parsing, we mainly utilize two well-known methods: constituency parsing or dependency parsing. Constituency parsing would involve breaking down our sentence into various hierarchical structures like noun phrases, verb phrases or prepositional phrases. Examining how these pieces connect then builds our chosen tree-like pattern capturing grammatical attributes throughout its branches.

Dependency parsing follows another methodology by focusing on analyzing the various relationship connections between spoken words instead. For example, analyzing subject-verb and object-verb relations within given sentences. From this method, we obtain a directed graph where each word represents its node while edges reveal connections between them.

In many different applications, Parsing has numerous uses i.e., language translation or text-to-speech conversion projects. In Machine Translation, Parsing can ensure that correct grammar choices are used for all relevant translations to create fluent sentences throughout programs.

If we take the sentence, "The cat sat on the mat," several parser types parse it when observed from multiple angles and perspectives; with constituency we evaluate (S (NP (DT The) (NN cat)) (VP (VBD sat) (PP (IN on) (NP(DT The) (NN mat))))). When relying on dependency evaluation, we use data- root(ROOT-0,sat-3), det(cat-2,The-1), nsubj(sat-3)(cat2); case(mat6,on4), det(mat6,the5); prep(sat3,on4),pobj(on4.mat6). Thus information producers can determine necessary components of speech and their primary relations from within the digital format of discourse information at hand.

Named Entity Recognition (NER) is a vital element of any complex Natural Language Processing system aimed at identifying and extracting specific Named Entities embedded within unstructured text inputs. Some common examples include Personal/User names (People), Business/Product names (Organizations), Location Names (Places) amongst other examples. The primary goal behind integrating this technique is mainly responsible for automating object identification tasks on large scale unstructured text datasets quickly and accurately. Once deployed, this technology utilizes statistical modeling combined with machine learning algorithms that analyze raw input data pieces through various multi-layered pipelines tasks. Surface patterns relating to named entities are detected then tagged accordingly. Once these steps are completed, highlighting Named Entities embedded within text data becomes perfectly possible; for instance if you come across the text, "Jack works for IBM Corporation," machines will automatically identify Jack as a person and IBM Corporation as an organization.

To perform this NER based analysis on text data at scale, machine learning models are trained on pre-existing annotated datasets that humans have created earlier. This preparation step ensures that machines auto capture similar named entities in new uncategorized records without requiring any human intervention when classifying identical entities/data points associating with relevant categories automatically. In summary, when dealing with unstructured textual data sources such as emails or chatbot logs sophisticated preprocessing techniques—like NER—serve as fundamental methods to transform dirty textual data into meaningful outcomes for further processing. In most cases, these preprocessors act as building blocks for complex entity classification categorizations done at scale. By adding advanced machine learning techniques alongside these basic but critical preparatory approaches, they serve critical purposes while enabling high level Natural Language Processing workflows to thrive.

Feature engineering and representation for NLP: bag-of-words, n-grams, term frequency-inverse document frequency (TF-IDF), word embeddings, and contextualized embeddings

In NLP, Bag of Words Representation is used to convert text data into machine readable format. This means creating numerical vectors that indicate how often each unique word appears within a given document—without focusing on positioning or order.

Rather, we treat each individual sentence like they're unorganized bags of words.

This technique starts by pulling all the unique words out of various documents before creating its specific vector reflecting one particular document with posted values standing in place of frequency pertaining to each individual letter amongst all resulting words.

Let's look at Example A vs Example B: "The cat chased the mouse." and "The dog chased the cat." Once we exclude their non-unique data elements, we can create numerical vectors that showcase how many times exact letters come up with Sentence A being [1, 1 1 1, 0] and sentence B garnering [1, 1 1, 0 1]. The numbering products stand for the certain number of times ``the""cat,""chased" appeared, among other corresponding items.

However, it's essential to know that this method comes with limitations such as ignoring word order plus context elements such as sarcasm or negation. Despite the potential uncertainties, utilizing this process works wonders for NLP applications including sentiment analysis and text classification.

Another valuable tool in NLP generating numerical representations of texts is called N gram Representation. When studying written material, it's common to examine sequences of words known as n-grams. For example, if we consider the sentence, "The quick brown fox jumps over the lazy dog," this phrase produces 2-grams known as bigrams like "the quick" and "lazy dog." By using phrases this way in Natural Language Processing (NLP), we can classify content based on assigned tags or anticipate future word sequences through computing each unique n-gram's frequency within given texts.

By counting each individual occurrence of certain terms in reviews about movies, for example, machines can assimilate comments into broader categories like positive or negative feedback. Here, specific linguistic tendencies like similar structures and phraseology appear frequently according to those predictive models thus generated informally from various data drawn from many reviews. Within NLP lies another numerical measurement technique called Term Frequency-Inverse Document Frequency (TF-IDF), which enables representation of documents with numerical values. The purpose of TF IDF is to assess the significance of a word in a document or collection of documents by comparing its frequency in the document or corpus with its frequency in other documents. TF IDF is founded on two measures: term frequency (TF) and inverse document frequency (IDF).

Term Frequency (TF) signifies the number of times a word appears in a specific document. The greater the term frequency, the more noteworthy it is for that particular document. Inverse Document Frequency (IDF) conveys how unique a word is across all documents within the corpus. Words that frequently appear throughout all documents are not considered significant for any particular document thus possessing a lower IDF score. On the contrary, words rarely seen throughout all documents are deemed to be more important to that specific document receiving higher IDF scores. The TF IDF score for each word is calculated by multiplying its term frequency with its inverse document frequency. Therefore, if a word appears frequently in any particular document but rarely within other documents, it will obtain a high TF IDF score indicating that it holds vital importance for that given text. For instance, there are three different types of documents within our corpus, and we intend to ascertain their respective TF IDF scores relating to the term "dog." In that case, we can use this method in each document as follows:

- Document 1: 3
- Document 2: 1
- Document 3: 0

The inverse document frequency of "dog" is calculated as follows:

- IDF("dog") = log (total number of documents / number of documents containing "dog")
- IDF("dog") = log (3 / 2)
- IDF("dog") = 0.176

The TF-IDF score of "dog" in each document is calculated by multiplying the term frequency by the inverse document frequency:

- Document 1: 3 x 0.176 = 0.528
- Document 2: 1 x 0.176 = 0.176
- Document 3: 0

Thus, we can see that "dog" is most important in Document 1, somewhat important in Document 2, and not important at all in Document 3.

The TF-IDF technique is widely used in text classification, information retrieval, and other NLP applications to identify the importance of words in a document or corpus. It helps in extracting meaningful insights from text data by identifying important terms and filtering out the noise.

Natural language processing (NLP) is an exciting field where techniques like embedding words using high dimensional vectors have been developed to help computers achieve better understanding and predictive power when working with human expressions encoded in text form. The embedding technique involves mapping each semantic aspect of a given vocabulary item onto specific dimensions within a continuous vector space allowing similar vocabulary items to acquire similar vector representations. This ultimately promotes improved contextual insights into sentence structure and meaning.

A commonly used algorithm for achieving such mappings is Word2Vec: it trains neural networks over large corpuses of text based data until these networks become able to predict likely outcomes based on contextual data. This generates sufficient information with which to build vector representations by which synonymic and related words can be identified and processed efficiently.

Through the use of embedded word techniques, computers have gained better natural language processing capabilities—an advancement that has enabled machine translation, sentiment analysis, text classification and more.

Machine learning algorithms for NLP

Clustering techniques in Natural Language Processing (NLP) form an indispensable part of managing vast Textual databases by grouping similar data points together. This technique allows us to separate specific aspects discussed within broad categories e.g., customer reviews about multiple aspects of a specific product analyzed through clustering can identify concerns about the product's durability or ease-of-use quickly. Several algorithms are deployed via high-level layering into subclusters by comparing lexical features like word frequency or various embeddings to facilitate efficient categorization across many types of datasets.

Applications within varied NLP projects include topic modeling for large corpora research such as sorting News articles within topical domains like politics, sports, and entertainment, etc. Sentiment analysis benefits from this technique, grouping similar feedback together around specific product keywords that allow efficient evaluation over vast datasets. Additionally, clustering helps Information retrieval processes involving search queries and documents similarity testing effectively by organizing retrieved datasets into manageable subsets related to querying keywords. Making search engine results more organized and user-friendly can be achieved by using sequence labeling techniques through NLP.

With this strategy, all associated data is classified accordingly by assigning individual context-based labels for every item within sequential data records using classified algorithms like Hidden Markov Models (HMMs), Conditional Random Fields (CRFs) or Recurrent Neural Networks (RNNs). One example could include describing whether an element belongs to people's names, locations or organizations while analyzing contextual information from its surrounding counterparts effectively. Composed of two critical components: transition model and emission models, HMM plays an essential role in natural language processing.

While transition models describe probabilities for changing from one implicit state to another, emission models determine probabilities for creating visible variables out of those states. To train a desirable Hidden Markov Model instance, you have to first estimate parameters simultaneously for both elements using labeled data points as reference.

Viterbi Algorithm comes in handy when trying to identify exclusive Hidden States among your observation sequence; Baum-Welch Algorithm will provide assistance with parameter calculation based on pre-trained instances compared against your expected functions. It's also critically important that we understand other useful Machine Learning Alternatives like Regular Expressions and unigram/bigram language models since selecting an appropriate method is sure based on circumstances peculiar to task specifications.

Deep learning models for NLP: convolutional neural networks (CNNs), recurrent neural networks (RNNs), long short-term memory (LSTM) networks.

Deep Learning has been a game-changer for Natural Language Processing (NLP). With advanced deep learning models capable of recognizing complicated patterns and dependencies among vast amounts of textual data, we can detect dramatic improvement across a wide variety of previously formidable tasks.

Notably, Convolutional Neural Networks(CNNs), Recurrent Neural Networks(RNNs), and Long Short-Term Memory(LSTM) networks type Deep Learning models have achieved significant success in the Natural Language Processing(NLP) field. CNNs are frequently used as they can extract features automatically by applying filters over input sequences; an essential difference from the earlier manual feature engineering technique.

These convolution filters typically consist of many small weight matrices that slide repeatedly over an input sequence and use element-wise multiplication and summation. Consequently, the model becomes better suited for grasping different complex linguistic attributes embedded in structured works like syntactic structures, n-grams or parts-of-speech. In addition to these useful functions, multiple filters with varying sizes are employed to allow flexibility in measuring abstract relationships and representations given substantial textual information segments analyzed. Later, at the pooling stage, minimum/maximum computations of values within small windows ensure better resiliency against inconsistent language use and the presence of user generated slang or idiomatic diction.

Finally, fully connected layers are available for identification/classification purposes thus enabling extraction of features across boundaries such as news classifications, whether it be sports-related or politics-affiliated topics etc. Conventional neural networks experience challenges handling variable length sequences such as text—this is where Recurrent Neural Networks (RNNs) are useful. Unlike standard neural networks that process fixed input sizes, RNNs efficiently manage inputs of variable lengths by retaining an internal state or "memory" of all preceding input they've processed. This aids in providing context for all words in sequences rather than analyzing each one independently.

An essential role played by RNNs is being used for language modeling tasks—predicting probabilities of next words given past word contexts in sequences—vital applications include machine translation and text generation.

Additionally, training an RNN on massive labeled datasets like customers' reviews can be useful for sentiment analysis classification tasks.

To remedy traditional recurrent neural network complications such as vanishing gradient issues faced during training procedures, alternative streak diagrams were invented—Long Short-Term Memory (LSTM), Gated Recurrent Unit (GRU). LSTM implements modifications, including an extra memory cell and 3 gates: input gate, forget gate, and output gate that greatly benefit Natural Language Processing tasks like language translation or speech recognition.

The operation of input gates controls how much new information flows into the cell state; on the other hand, gates limits which data gets deleted from it. Output gates regulate how much information emanates from this location within LSTM networks that can recollect previous details effectively because of their Memory Cell property. It serves as a conveyor belt moving data between every timestamp, with added features for adding or reducing facts en route.

To illustrate this concept's application efficiently within natural language processing (NLP), imagine predicting upcoming words in a sequence accurately using traditional Recurrent Neural Networks (RNN). They may show reduced accuracy when dealing with intricate inter-dependencies among various verbal expressions present in text data sets. Contrastingly, employing LSTMs enables us to incorporate knowledge gained during previous time-steps and makes predictions more accurate concerning textual data. Empowering developers in mastering natural language processing tasks are models such as CNNs, RNNs, LSTMs, and transformers. Choosing an appropriate algorithm depends on specifics of an assignment and characteristics presented by textual input.

We must comprehend the strengths or weaknesses of each model for effective development rates in producing accurate NLP systems.

Advanced topics in NLP: sentiment analysis, topic modeling, machine translation, question answering, summarization, and dialog systems.

Natural Language Processing (NLP) has made significant strides since inception with considerable recent progress achieved allowing application of advanced topics. These topics include sentiment analysis, topic modeling, dialog systems, question answering, summarization and machine translation across numerous fields ranging from finance, marketing to healthcare and education.

Sentiment Analysis involves analyzing and understanding emotional tone embedded within texts with aims of scoring specific sentiments expressed whether these are Positive, Negative or Neutral by implementing varied techniques combined with algorithms aimed at spotting such sentiments. Using such techniques enables companies to note customers' feelings/attitude toward their products by accessing reviews/comments on social media platforms.

Sentiment Analysis techniques frequently involve usage of rule-based systems, machine learning/algorithms and deep learning models depending on varied features for recognition purposes, be it word frequency/part-of-speech tags/contextual detail. In-depth insights gained from such analyses are pivotal for enhancing brand reputation measures/easing customer interactions besides other business applications.

Topic modeling is yet another NLP technique applied to identify hidden themes via analyzing recurring patterns/themes present within datasets; some widely used topic modeling algorithm is the Latent Dirichlet Allocation (LDA). The original article describes LDA as a probabilistic model used for representing documents as combinations of subjects assigning probabilities indicating its relevance among particular themes. The model proves helpful when identifying common topics such as politics, sports, entertainment or technology in news collections and correlating them with similar articles based on what's being talked about most often using LDA's probability assignments as indicators.

Corporations use it extensively during customer review analysis procedures where discovering consumers' commonly voiced opinions is important so they could effectively improve products/service offered by addressing specific problematic areas. For example, electronic product manufacturers use topic modeling to gain insight into what customers' opinions are of their offerings in multiple countries. There are several machine-translating tools available today, and Microsoft Translator is one such example utilizing statistical methods to translate both text and speech assets across multiple languages.

It's an essential aspect within international business and diplomatic areas where accurate communication overcomes barriers created by language constraints. Another area where NLP has gained popularity is through Question Answering (QA), a means of creating computer programs that mimic the human process when handling questions posed in natural language.
 An application of NLP: dialog systems (also known as chatbots) replicate human conversation between users and machines by utilizing natural language processing (NLP). These intelligent programs are integrated in customer service interaction platforms where serving clients is conducted around the clock; assist with daily schedules; reminders etc.

Personal assistant apps are heavily reliant on rule-based instructions by retrieving responses routed from a pre-existing set of data answering specific questions posed by the user with specified rules. On the other hand, generative systems work with reinforcement learning techniques generating responses that are more natural.

Evaluations of these systems should be based on user feedback, response quality measured by the F1 score and task completion rates. In recent years, there has been considerable progress in cutting-edge topics such as sentiment analysis topic, modeling machine translation, question answering summarization, and dialog systems. For constructing successful NLP systems, understanding the benefits and limitations of each method is essential.

Transformers

The focus gadget performs an indispensable role in developing deep learning structures dedicated to natural language processing (NLP). It lets developing structures pay detailed attention to particular inputs when forecasting or processing data.

While Jacques Pitrat proposed an early attention system back in the 1970s that identified important pieces within input settings, it wasn't until far later that architectures equipped with high-class focus systems gained prominence. The fast advancement witnessed around this time saw researchers at Dzmitry Bahdanau et al. introduce an attention mechanism to facilitate machine translation in 2014, which achieved desirable results.

Research undertaken by Vaswani et al. revolutionized the focus device's functionality by allowing models to selectively attend to various parts of the input sequence during encoding without requiring any explicit linkage. Consequently, this breakthrough prompted the introduction of transformer models into NLP via the paper, "Attention Is All You Need."

Prior methods such as recurrent neural networks (RNNs) and convolutional neural networks (CNNs) were ineffective when managing lengthy text sequences, creating obstructions that impacted training outcomes negatively. Attention mechanisms removed positioning restrictions between two entities in a sequence and enabled crucial improvements such as state-of-the-art machine translations performance through transformer models.

Attention mechanisms have become commonplace among many state-of-the-art natural language processing (NLP) models such as BERT, GPT-2 and RoBERTa over recent years. Attention works by utilizing different types of attention mechanisms, for instance, cross-attention and multi-head attention crucial for improving performance metrics across various NLP tasks like sentiment analysis, machine translation and language modeling.

Transformers are unique neural network architectures for NLP introduced first in 2017 named after their ability to transform input sequences into desired formats seamlessly. They surpass previous NLP models due to their exceptional competence at capturing long-range dependencies prevalent with text completion algorithms that were extremely limiting factors for older models. Self-attention is fundamental in transformers' operation where every word assigns relative weights based on contextual relevance learned during training.

Language modeling represents a critical application of transformer technology within natural language processing (NLP). Language modeling capabilities allow transformational technologies like transformers to predict subsequent phrases using preceding contexts. Consider an example where a user types, "I went down ___," following the phrase "movies last night." The prompt would then suggest most likely choices like "store" or "park", highlighting the transformer's critical reliance on context and self-attention while making predictions.

In summary, transformers come with encoder-decoder models—the encoder generates hidden representations from inputs, while decoders use these hidden representations sequentially to generate output sequences. The transformer's encoder and decoder layers employ several self-attention and feed-forward units, each involved in advanced computations for a broad range of Natural Language Processing tasks.

During self-attention processing, traditional embeddings created from vector-formatted words within an input sequence, undergo a refinement that measures each component's significance relevant to other words or components within that sequence.

Using three distinct matrices—the query, key, and value—trained weights are utilized simultaneously throughout the transformation process to create attention scores for all possible pairs across all words within the sequence.

The newly refined set replaces input embeddings subsequently assisted by feed-forward operations before passing through additional units resulting eventually in more intricate correlations established between manipulated input-output sequences.

Thanks to their ability for supporting complex functions this way, transformers excel at facilitating many NLP-related tasks including machine translation endeavors when required. Here, they generate an equivalent output sentence based on inputs given initially by transforming one actual sentence from one language into its meaning transferred over into another language via prompted response generation.

OpenAI's Generative Pre-trained Transformer (GPT) serves as a beacon model of successful implementation of such architectures that are configured strategically through extensive pre-training on vast amounts of textual data to replicate human-like output during prompt-response experiments. Founded in 2015 by prominent tech industry leaders such as Elon Musk and Sam Altman, OpenAI is an artificial intelligence research lab focused on creating safe and effective AI solutions for addressing some of the most significant global challenges. The company's initial focus was creating specialized AI that could carry out various intellectual tasks similar to humans instead of pursuing Artificial General Intelligence (AGI).

Subsequently, OpenAI has achieved many milestones in natural language processing technology; among its most significant was launching the first version of Generative Pre-trained Transformer (GPT) model in 2017. This cutting-edge deep learning model uses transformer neural network architecture capable of processing vast amounts of text data to produce coherent and fluent text. This process results in NLP applications like chatbots or text classification becoming reliable and accurate across different domains.

The company continued refining GPT's models with GPT-2 or GPT-3 that bear improvements over their predecessors, which possess remarkable proficiency throughout various NLP tasks such as translation or summarization amongst others. Additionally, OpenAI has contributed to impact other advanced fields like robotics, computer vision, gaming expertise while collaborating with Microsoft resulting from more extensive accessibility toward their targeted audience via advancements made possible through joint ventures. This enables shared technological achievements across multiple domains to simultaneously form a broader community network involving interdisciplinary integration. This process serves many purposes across different generations using efficient innovation techniques despite controversies surrounding transparency that is lacking within the technical elements involved during developmental stages. Nonetheless, we can consider open Ai as one of the leaders driving AI research toward transformative possibilities.

The inception story behind GPT dates back to 2017 when Google introduced their groundbreaking "transformer" architecture through a research paper. Using attention mechanisms to process input data made this neural network methodology exceptionally useful for tackling NLP challenges. In 2018, OpenAI's team leveraged this transformer concept and built their first version of GPT by training it on an enormous corpus of textual data. This pre-training helped calibrate the system accurately toward generating coherent and plausible content.

Evaluation metrics and methods for NLP: perplexity, BLEU, ROUGE, and human evaluation.

To enhance natural language processing (NLP) systems' efficiency rate means we need to critically evaluate them periodically. Such assessments help researchers understand how effective their systems are running and where necessary changes may be required to improve functionality effectively. We utilize various methods such as accuracy rates, recall percentages assessment procedures like F1 score evaluations. These include confusing multi-word nonsense sentences or structured questions with definitive terms encompassing complex topics within complicated sentence structures.

One important metric for evaluating a language model's efficiency is perplexity. Perplexity looks at how effective the model is in predicting a sequence of words given previous content in the sentence. The calculated perplexity value then uses the inverse probability of a test set, normalized by the number of words in that set to calculate whether our system performs better in predicting next-word sequences or if further improvement is necessary.

Consider this hypothetical situation: suppose we designed an NLP system tailored to news articles and have just received our first set of test data focusing on topics such as machine translation or text classification tasks. Calculating the perplexity values with these tests can give insight into efficiency improvements necessary for said subjects.

Lastly, BLEU (Bilingual Evaluation Understudy), as a metric, helps establish machine-generated translations' quality compared to human-translated reference texts. BLEU aims to compare human translators' reference translations against machine-generated alternatives and ensure that our systems perform well by accurately capturing semantic meaning without errors. The BLUE score can be calculated by analyzing n grams in machine generated translations and comparing them to n grams in reference translations.

The number of matching n grams in the machine generated translation is compared to the total number of n grams to determine its precision. Similarly, recall is calculated by comparing the number of matching machine generated translation n grams with the total number of reference translation n grams. Finally the BLUE score is obtained by calculating the geometric mean between the precision and recall scores.

For example, let's say we have a machine-generated translation of a sentence: "The cat is on the mat." We also have three reference translations of the same sentence:

1. The cat is lying on the mat.
2. The cat is resting on the mat.
3. The cat is sitting on the mat.

We first break down the machine-generated translation and the reference translations into n-grams. Let's use bi-grams for this example, which means we will consider pairs of adjacent words. The machine-generated translation has the following bi-grams: "the cat", "cat is", "is on", "on the", and "the mat". We compare these bi-grams to the bi-grams in the reference translations and count the number of matches. Let's say we get the following counts:

- Bi-gram "the cat": 3 matches
- Bi-gram "cat is": 2 matches
- Bi-gram "is on": 1 match
- Bi-gram "on the": 0 matches
- Bi-gram "the mat": 3 matches

The precision for this machine-generated translation is (3+2+1+0+3)/5 = 2/5 = 0.4, and the recall is (3+2+1+0+3)/15 = 2/5 = 0.4. The BLUE score is the geometric mean of the precision and recall, which in this case is sqrt (0.4 * 0.4) = 0.4. A perfect translation would get a BLUE score of 1, while a completely random translation would get a score of 0.

Researchers in natural language processing often need rigorous ways to measure whether their machine-generated summaries contain all essential information presented in original source materials. Recall-Oriented Understudy for Gisting Evaluation (ROUGE), one of many evaluation tools available, offers a clear benchmark for assessing results produced by automatic summarizers. We use ROUGE by comparing machine output with human-made paraphrases and calculating overlap between them using either n-grams or categorical word sequences as benchmarks.

As it relates specifically to ROUGE, scoring is determined primarily through comparison with reference summaries based on recall calculations. This means that we assess how much of these original references are contained in our summarizations outputs. Different users may require varying values of 'n' depending on their specific needs for detail level and matching requirements during evaluation processes.

With objective measures like ROUGE at our disposal, we can better gauge development successes; nevertheless utilizing supplementary evaluation mechanisms such as human judgments allows us deeper insight across a wider range of perspectives surrounding text summarization sophistication and effectiveness overall. The evaluation of NLP models or systems is a significant component in Natural Language Processing (NLP) that requires presenting generated outputs by these systems to human evaluators who then rate or provide feedback based on fluency accuracy relevance coherence amongst other features. Its main objective ensures that these NLP models produce high-quality outputs that fulfill user requirements.

Commonly used applications of human creations include Machine Translation (MT), Summarization Question-Answering (QA) among others. For instance, MT tasks require presenting text in two different languages alongside a translation, then evaluated based on factors such as fluency accuracy and general quality.

Despite being subjective due to varying opinions from different evaluators, making use of multiple evaluators can give more reliable results.

Accurate metrics such as Precision Recall F1 score perplexity BLEU ROUGE etc. and human evaluations remain valuable methods used in assessing the effectiveness of NLP systems; however; it's highly important to interpret those results within the context of their specific applications accordingly.

Applications of NLP in various domains: social media, healthcare, finance, education, law, and customer service.

Natural Language Processing (NLP) has been adopted across different domains due to its potent capabilities when automating tasks exclusively reliant on human intelligence in recent years. The efficient incorporation of these automations into customer service solutions within healthcare systems; Finance sector; educational institutions; among other industries has significantly impacted operational performance toward meeting rapidly growing needs driven by increasingly diverse personalities.

Social media utilization has revolutionized communication, turning it into what can be described as a data mine filled with value capable of being mined for actionable insights through natural language processing analysis. Through NLP derived insights, substantial value addition that enhances overall efficiency for companies can be obtained by analyzing people's sentiments and moods provided by social media, which determine consumers' perceptions for brand quality.

Additionally, NLP empowered Chatbots guarantee customers with immediate resolution of routine inquiries thus saving valuable time among customer representatives leading to better feedback toward improved customer satisfaction ratings. Social media monitoring results in real-time and relevant data on brand mentions. Its impact as well as critical sentiment trends within the conversation providing valuable insights to enable more informed decisions geared toward marketing campaigns optimization.

Ultimately Personalized advertising facilitated through Natural Language Processing Analytics ensures that adverts have personalized messaging aligned with individual users' habits and interests while offering balance in the user-advertising relationship. Here, the ads content delivered feels less intrusive while reflecting on both business revenue optimization objectives.

NLP software has rapidly become an essential tool for healthcare providers looking for ways to improve patient care while adhering closely to regulations. By analyzing free text clinical notes that would typically go unexamined under traditional approaches, NLP identifies important information such as diagnoses or prescribing data—deepening clinicians' understanding of patient history while informing more informed decision making processes.

In finance, sentiment analysis using NLP software offers a way for financial institutions to predict market trends by analyzing social media posts or other public sources of text information. This tool supports everything from assessing customer demand for particular investments or product offerings to facilitating fraud detection measures that allow for proactive risk management practices. In both sectors, the long term potential benefits are wide reaching and impactful—paving the way for future technological advances in these areas. Investors have access to insights that can inform sound investment choices by leveraging the source this information's powerful analytical capabilities effectively provided herein regarding buying or selling stocks intelligently. An additional application area where Natural Language Processing (NLP) is invaluable relates toward fraud detection. With NLP technology, it becomes possible to scrutinize customer complaints, online reviews—along with other data sources—for any fraudulent activities that require financial institutions to take appropriate corrective measures.

Investors can use trading algorithms that leverage innovative NLP-powered analytics to study factors such as financial reports or news articles in deciding trades based on specific criteria. This utilization of technology has the potential to enable informed decisions amongst investors and bring about better returns on investments.

Leveraging NLP-powered chatbots in the finance industry's customer service realm makes it effortless for customers and staff alike. The system intelligently sorts queries past its algorithms, establishing personalized responses tailored toward customers' specific needs, quickly providing solutions for their queries.

NLP is also critical for risk assessment purposes within the financial industry's lending services environment.

Through analyzing credit reports as well as language patterns present in loan applications-data—financial institutions using these analytical tools can make more informed decisions when rendering loans.

Education domain stakeholders similarly utilize NLP technology effectively. Teachers benefit from a better understanding of students' learning needs through engaging communication with them; they can offer optimal support tailored explicitly toward those individual student needs. Through these particular tools' utilization, language assessments become efficient; writing skills developments are monitored to provide academic improvements continuously.

Moreover, students receive personalized learning experiences from their unique profiles' construction using behavioral and language data analyzed by powerful NLP tools used sensibly within this educational space.

Finally, recommending educational content such as reading materials or videos, which are increasingly compatible with individual student's abilities is made feasible via such technologies deployed here. This is done through sophisticated analytics provided by advanced Natural Language Processing (NLP) systems making education directly beneficial to more learners globally. For students who struggle with traditional methods of learning, NLP technology can be a game changer as it gives them the flexibility to work at their own pace according to individualized needs.

Similarly, advantageous features also exist for educators; an NLP chatbot is perfect for improving communication between teachers and learners by giving quick responses on common inquiries like assignment instructions or exam schedules—saving both parties a significant amount of time! Moreover when used appropriately across platforms like email or social media posts, NLP analytics algorithms may detect patterns of bullying or harassment allowing swift action to be taken.

NLP has proven useful in improving legal compliance as well. Using NLP's analytical capabilities on regulatory documentation regarding compliance can aid legal practitioners to identify potential areas of concern. In turn, this helps ensure client conformity with pertinent laws/regulations as mandated by regulators governing their operations. Furthermore, it also helps educators enhance their teaching processes through analysis of texts from case-law jurisprudence available across a variety of textbooks on legislation/policy relevant to their field.

It is vital to note that beyond improving outcomes within law using its many facets, an array of other domains have benefited from NLP across healthcare services, finance, social conversations, etc. Large datasets which would once take hours or days are now streamlined for fast queries. Automation now means accuracy, ensuring unparalleled levels of assurance both for businesses and customers with better interactions between them being noted resulting in greater convenience. Our reliance on increasingly sophisticated algorithms particularly within natural language processing highlights how important this technology continues to be with advances constantly occurring over time where changes are inevitable.

Future directions of NLP research and development: multimodal and cross-lingual NLP, explainable and interactive NLP, and NLP for emerging technologies such as augmented reality, virtual reality, and the Internet of Things.

As researchers work diligently on finding novel avenues for Natural Language Processing (NLP), this field keeps expanding rapidly. This section explains the future directions of NLP research by highlighting multimodal and cross lingual NLP along with explainable and interactive NLP.

Multimodal communication involves considering numerous forms of inputs/outputs such as images, videos alongside speech/text, which improves the accuracy in conveying the intended message better since humans use additional cues like intonation and body gestures too while communicating. To illustrate further—you may spot various cooks searching for delicious recipes online using both visual aids such as photographs/video and text. In this case, image recognition technology with NLP can recognize ingredients and recipe steps mentioned in pictures.

Another fascinating example is how virtual assistants like Siri or Alexa utilize a combination of speech recognition, natural language processing and text to speech synthesis, which enables these systems to deliver a more seamless and natural user experience. The next direction that NLP technology is heading in involves the cross lingual approach which implies comprehending various languages. Every language has its unique sentence structures and grammar rules; however lots of languages share similar meanings or ideas as well. Hence, processing multiple languages accurately requires different techniques to ensure natural communication while including multiple languages. Explainable and interactive NLP pertains to natural language processing's ability to clarify its output while actively engaging users for a tailored experience.

Essentially, using NLP warrants adequate lucidity on how the system arrives at recommendations or conclusions. Explainable NLP relates to delivering concise explanations that are easy-to-understand regarding how the system settles upon its results while transparently disclosing its rationale—helping foster trustworthiness among users who rely on that information for decision-making purposes. Interactive NLP permits users to partake actively in enhancing a system's relevance by offering their feedback. This accurately reflects their question or intent context awareness betterment being crucial case-on-point for customer service chatbots where interactive feedback would help customers provide clarity about their questions or issues that require resolution. The Internet of Things (IoT) pertains to a set up where physical devices such as buildings or vehicles embed sensors and software allowing them to connect among themselves and exchange data without human intervention.

When these IoT gadgets are used within NLP settings, particularly voice recognition together with natural language understanding, this allows smart speakers like Amazon's Echo or Google Home's integration of IoT technology recognition into responding via voice commands effortlessly.

Another example takes form through bilingual tourists who require translation services where IoT-enabled devices use NLP algorithms providing accurate instant translations in various contexts like business meetings or while visiting a particular tourist destination. In essence, there lies an optimistic horizon for future innovation in research pertaining to natural language processing—one that is teeming with potential excitement. Currently at the forefront of spirited exploration within this field are: multimodal / cross-sectional linguistic processing; explicative / interactive natural language programming; plus developments amidst emerging technological landscapes inclusive of AR/VR tech as well as IoT-oriented applications. As these various platforms continue their headway toward advancement, it's plausible to expect new-fangled ways/uses uncovered via integration with cutting edge linguistic software.

GENERATIVE AI

Introduction to Generative AI: This section provides an overview of what Generative AI is and how it differs from other types of AI, such as discriminative models. It also explains the basic concepts and terminology of Generative AI, such as probability distributions, latent spaces, and autoencoders.

Generative AI falls under the larger scope of artificial intelligence where its primary focus centers around creating unseen but similar outputs using machine learning rather than categorizing based on existing datasets; because it focuses on generative pattern algorithms rather than discriminative.

The practical applications for this kind of technology are numerous; generated artwork, unique musical structures or novel writing style solutions are only some examples on the medium scale while output possibilities range widely—including video, sound or imagery. The overall endgame for investigators in this field is for machine-generated outputs that match human skill levels at worst case scenarios—possibly even surpassing our creative faculties someday.

GPT-3 demonstrates how far Generative AI has come in text generation tasks as it creates very human-like statements based on pre-existing text-data histories discovered via deep learning techniques.

The fact remains, however, that Generative AI outputs must be questioned for ethical considerations such as bias and fake news. The constant evaluation of these AI systems during their lifetime is imperative to ensure ethical compliance on all levels.

Autoencoders offer a specific deep learning technique—neural networks—which mainly concerns itself with compressing current data before decomposing it into a new outcome. The basics of autoencoders are composed of two sections: the Encoder responsible for reducing input data such as images or text into compressed latent codes capturing essential features; and the Decoder tasked with reconstructing the original input from its latent code.

An autoencoder model extracts latent code, which produces a reconstructed output that mirrors original input as closely as feasible. Such models have a number of applications in generative tasks such as text and image generation. For instance, the autoencoder, trained on facial data, can learn to create images/models that resemble faces from the provided data set. Similarly, if provided with pieces of written text as input data corpus, it trains itself to generate new comprehensible topics resembling the patterns and structures of its original material. Besides these tasks, other utilization of autoencoders include feature extraction or anomaly detection, which is useful in an unsupervised learning environment.

Generative adversarial networks or GANs are advanced deep learning models utilized most often with generative AI. These networks are composed of two neural networks: a generator network and a discriminator network. The generator network processes random noise inputs producing fake samples similar to authentic ones while the discriminator network learns to differentiate between fake and real samples during training periods. This is done through different datasets analyzing for similarities or dissimilarities between them, both respective types, before reaching maturity level after feedback loops resulting in improved processing across domains. As training progresses over time by competition between these two networks concerning their goals, the generator aims toward producing increasingly realistic outputs that trick or confuse beyond its opponent's discerning capability. Discriminator aims at correctly identifying such artificially generated fakes that may be too tricky otherwise even for humans' naked eyes visually inspecting them separately making judgments accordingly.

Once fully trained, GAN-generated samples achieved remarkable success in multiple fields such as image/video synthesis and natural language generation (using machine/deep learning techniques). For example, GANs applied on celebrity facial dataset models virtual celebrities, closely resemble those public figures recognized by people worldwide. These virtual celebrities display excellent capabilities within this field demonstrating how far AI has advanced over recent years doing artificial generational authentication. These have become almost similar yet distinctively different from human-created content providing excellent results close enough not only to deceive but close enough mostly mimicking realistic nature indicating exceptional AI functionality through GAN model versatility.

In order to create a new sentence or paragraph structures with similar style and content to pre-existing ones, GANs can be utilized after being trained on relevant text corpora. When we refer to probability distribution in generative AI context, it represents likelihood assessments for various potential outcomes in sets or combinations events. Generative AI relies heavily on probability distribution modeling; isolating distinct properties from other factors—this helps it produce individualized outcomes based on existing samples that follow comparable patterns. For instance, should we want realistic depictions containing certain aspects, like cats' fur color or ear shape, than available tools that help us take note of all parameters involved.

Gaussian or Bernoulli methods are some examples used often for generating continuous/binary types. Expectedly then varies between applications sought by type: another selection choice would be the Poisson method which counts occurrences in samples. Conclusively, one can assess that this groundbreaking technique of creational AI has far-reaching potential across different fields. It aids applications including creative arts, natural language processing and computer vision.

Types of Generative AI models: This section delves deeper into the different types of Generative AI models, such as Generative Adversarial Networks (GANs), Variational Autoencoders (VAEs), and autoregressive models. It explains the differences between these models, their strengths and weaknesses, and the types of problems they are best suited for.

Generative Adversarial Networks (GANs)

The technical methodology developed within Generative Adversarial Networks or GANs employs adversarial training—a form of deep learning algorithmic framework. The approach engenders competition between two neural networks with disparate responsibilities: one whose work sets out to generate images (generator), while the other checks whether they are verifiable as plausible naturally occurring examples (discriminator).

The process continues through multiple rounds of intense iteration between sampling creation and analysis by discrimination performance metric called loss calculation algorithm over time. The generator can be optimized to create more realistic images appreciated by its counterpart discriminator network, which will judge candidate representatives for plausibility. As this calibrated interplay generates new samples that successfully pass off for an actual example, we can say our GAN has "won". Generative Adversarial Networks (GANs) are promising models that have shown remarkable results in generating high-quality samples such as images and text. However, developing GANs involves several challenges such as mode collapse wherein only a few kinds of outputs are generated while ignoring other possibilities. In addition to this issue, vanishing gradients during training may hinder learning effectiveness, which makes achieving convergence increasingly challenging.

Additionally, unstable training conditions due to imbalance between the generator and discriminator come up as concerns too; especially in evaluating performance because there are no explicit objectives that are optimized with any known clarity or agreed upon consistency measures (Metrics like Inception Score and Fréchet Inception Distance exist but aren't foolproof).

Moreover, amidst these obstacles faced by developers lie those relating data availability—large amounts needed—and computational requirements difficulty; making it understandably tough for certain domains or individual uses such as limited computing resources. With GAN's undeniable ability to produce incredibly realistic and high-quality outputs like images, videos and audio, it's easy to see why they're popularly celebrated among artificial intelligence models today. They can help create completely new designs unseen before thus making them applicable in vast arrays of fields including but not limited to art and entertainment.

Another edge held by GANS is its collaborative eligibility; now artists', writers' works are modeled on various creative possibilities allowing room to select from an abundance of ideas—without losing quality thanks to data learned during its training processes. In the realm of generative AI techniques like GANs, VAEs stand out for their versatility in creating varied data points with widespread applications from compression to image generation.

Variational Autoencoders (VAEs)

Imagine this scenario: You're an artist embarking on your next project but unsure what path to choose creatively? Here enters the Variational Autoencoder (VAE), which helps generate ideas by learning subtleties of existing artworks through analysis of large datasets.

With this knowledge base established in a simplified version called latent space, representation captures essential components of original images via pattern analysis by the VAE algorithmic model. Random inputs provided subsequently convert into compressed latent functional spaces with crucial information on painting style rather than individual painting complexities.

As an artist using varying continuous sets among these crucial variables creates original pieces with distinctive styles, which is evident from different personal artistic sensibilities.

This is what gives us the power using these VAEs in the world of generative AI: diverse creative possibilities and new artistic developments across a wide range of fields. The path ahead includes overcoming challenges like training the VAE model effectively, creating quality output-generated artwork, and pushing limits to increase algorithmic refinement. VAEs present their own share of obstacles despite their technological prowess like any other invention. When it comes to generating novel and diverse data while still maintaining coherence and consistency, the balancing act becomes critical.

In this context, we are also concerned with finding harmony between encoding-decoding processes via mapping onto a lower-dimensional latent space that makes generating fresh material simple but can prove challenging in capturing more nuanced details surrounding the original information.

Hyperparameter selection poses yet another challenge requiring deep thought on dimensionality choices including which networks have how many layers alongside optimal learning rates; ensuring quality outcomes requires significant time investment combined with considerable computational cost. A separate issue arises when looking into evaluating model performance regarding VAEs. Conventional metrics are inadequate for VAE evaluations as opposed to other generative models. The commonly used ELBO (Evidence Lower Bound) may not offer a proper reflection of generated data quality concerning these models.

Moreover, Variational Autoencoders often face mode collapse where few modes from the dataset distribution get generated exclusively over others via the model's deep learning processes; this leads to repetitive or lackluster results.

Autoregressive Models

Using autoregressive models is a prevalent method in AI when it comes to generating sequences of data because it's easy yet powerful. Consider predicting what word comes next in a sentence; one way is by studying all preceding words and using that knowledge as guidance for your prediction regarding upcoming words in this content. These predictive algorithms work in a similar manner; however, instead of considering only language constructs like words or sentences, they are designed for any sequential dataset (image/ audio/ or text).

For instance, let's consider recreating handwritten digits using an autoregressive model—pixel-by-pixel guessing based on past pixels would create such results efficiently. Behind each outcome are precise calculations made by functions making use of mathematical operations called Neural Networks that learn from patterns via previous rounds while likewise providing outputs amounting to probability distributions from which data is then predicted.

With an iterative approach where parameters are adjusted during training, maximum likelihood estimation is employed to train autoregressive models. Here, the model is continually optimized to maximize the chances of producing accurate data following a prior sequence of information. Autoregressive Models showcase their flexibility by working with diverse datasets such as audio, images or text. They also have conditional capabilities ensuring data-based predictions that can be customized as input requirements guiding the predictive algorithm. Autoregressive models are frequently relied upon when it comes to generative AI since they predict the subsequent value by analyzing precedents.

An example use case could be generating cat images through an autoregressive model trained via certain breed inputs. Even so, such models come with a set of significant constraints.

Firstly, there is computational expense, which predominantly emerges from having to consider all preceding sequences up until that point for every single element within a sequence due to dependence on previous values. As this requires storing massive amounts of data into memory regularly for every computation made, leading both inference and training processes slow and computationally expensive.

Besides computational cost, another issue arises regarding modeling long-term dependencies within each sequence using this method. This is because it can solely base future predictions upon past observations leading to loss of essential correlations among input variables resulting in models producing less coherent and realistic output.

Lastly, exposure bias also is another factor to consider where the model during the testing phase produces its own predictions instead of ground truth sequences used during training. This may lead to increasing errors in output creation processes over time resulting in reduced quality output. In certain situations, an inconsistency between training and testing distributions can result in suboptimal produced outputs. Nonetheless, navigating challenges such as this has not stopped autoregressive models from being a powerful asset in generative AI. These models are capable of generating various media forms such as languages or music styles. When confronted with these difficulties, establishing a strong plan of action could allow for groundbreaking advancements with autoregressive models by innovative researchers.

Other types of generative AI

It's worth mentioning other kinds of generative AI models aside from GANs, VAEs, and autoregressive ones; practitioners employ these variations suitably for different purposes. Here are two examples;

Firstly, **Flow-Based Models** transform uncomplicated probability distributions into more intricate ones representing underlying characteristics present within datasets; this drives their ability to procure fresh copies of those same datasets that resemble the original but aren't identical copies. Essentially, Flow-Based Models consider dataset generation methods before producing new samples with similar traits. Suppose you aim to create a credible forest image; Flow-Based models can start with an image showcasing some trees, then gradually embellish these with details like leaves and shadows until the outcome is what resembles a genuine forest picture. Their key training process maximizes the likelihood of generating realistic images while utilizing negative log-likelihood minimization measures.

Flow-based models lend themselves to high-quality and diverse output types with relative ease in applications like text generation, speech synthesis, and image production. However, there are computational challenges associated with them; training on significant data sets can quickly take up computing time as they struggle to capture complex dependencies between different variables leading to poor performance on specific types of data.

Furthermore, **Generative Moment Matching Networks (GMMNs)**, another model type, produce near-realistic data through moment matching strategies. When it's hard to define precise probability distributions for datasets, using a GMMN can be useful—especially when you want to generate lifelike facial pictures from previously captured datasets containing genuine photos. Traditional generative models like Variational Autoencoders or Generative Adversarial Networks can have difficulties creating complex subtypes and diverse versions of face images, which necessitates going for non-traditional means like GMMN that involves finding parameters matching actual data patterns (such as mean or variance).

After parameters identification, they can be used to generate new and similar pictures from accurate mean and variance that resemble legit facial images. The learning algorithm of GMMN includes kernel mean matching—a process of finding suitable kernel functions that can represent the data structure. Parameters then get optimized based on these kernels until the generated samples' kernel moments resemble the real face image dataset. GMMN has an added advantage as it doesn't require explicit density estimation compared with other generative models like GANs, particularly in high-dimensional spaces. Moreover, their training processes are relatively more stable than other models.

However, these networks have issues capturing complex data distributions because they tend to struggle with datasets that have variability, resulting in generated outputs lacking sharpness and details when compared to PixelCNN. The complexity of this model rests within its expansive networked layers built from convolution techniques. At its simplest level, PixelCNN is given a monochromatic picture that it then builds incrementally from earlier pixel generations used as input. To create each new pixel, PixelCNN conditions its masked convolutions—an intelligent approach that restricts convolutional filters to earlier generated subsets of pixels. This tactic ensures sequential building while maintaining contextual consistency between prior and future pixels.

PixelCNN offers particular advantages as a fully autoregressive model such as generation of fine-detailed images with high variation. However, this style has costs in computational expense; generating high-quality images requires considerable storage and processing power, making large or detailed image generation more achievable. Nevertheless, PixelCNN successfully accomplished applications such as image production, completion, reduction of image noise and even simulation of realistic animal or human facial features. This effectively incorporates computer graphics or animation techniques.

Generative AI models are unique in their corresponding strengths and weaknesses; a GAN is capable of generating superior artificial images but difficult to train; a VAE maintains uncertainty within the data distribution but may produce blurry images; autoregressive models generate cohesive text yet can be slow in producing new data. Results rely on your project's specific context, and the trade-offs you consider most important including the quality improvement of generated data, the optimal speed for training/inference, and complexity. This chapter has been dedicated to showcasing how generative AI models can serve as a potent tool for generating new data similar to previous training sets with ease. The focus revolves around investigating the various forms—GANs, VAEs, autoregressive—which fall under this category's umbrella while assessing each model's unique features such as benefits, drawbacks along with practical implications across multiple domains.

Generative AI seeks to create new data sets that strongly resemble training datasets using various advanced techniques from artificial intelligence fields such as image and video generation, natural language processing, and drug discovery analysis domains.

Probabilistic modeling is one vital method employed within generative AI techniques for creating accurate probability estimation models for any particular set data provided by its creators. The power lies within building accurate new datasets without relying too much on pre-existing templates or patterns using this method's principle forms such as Gaussian mixture Model (GMM) or Markov chains.

With a GMM model applied to datasets containing flower images focused on capturing several gaussian distributions articulating different flower types present in the original group of training datasets, generating new images from randomly sampled learned distributions is very straightforward.

Markov chains bring dynamic modeling benefits to generative AI by using precedents like various music notes or even entire sentences as an earlier reference point for anticipating future sequences' rightful probabilities. Optimization is critical for generative AI models as it aims at finding model parameters leading to minimal objective function values. A mathematical algorithm known as an optimization algorithm drives generative artificial intelligence's functioning, helping it achieve its objectives by minimizing differences in output and actual data through parameter adjustment using gradient descent techniques. Nonetheless, optimizing these AI models proves difficult due to high dimensionality in their parameter spaces and complex objective functions. These models come with standard optimization algorithm limitations like gradient descent approaches, which make it difficult to find global minimums of loss functions universally negative. To overcome these issues researchers have developed specific optimization options suitable for generative AI models such as adaptive or adversarial algorithms while altering sample processes for more accurate output reflection across a wider range in random sampling methods or beam searches. Here, most likely, sequences that are iteratively chosen make predictions for higher-quality generated-data results reflecting original Data distribution.

A strategy favored among natural language programs such as text-generation or language-translation is Monte Carlo Sampling which garners superior outcomes emulating source-data distribution with precision when deployed during generative AI processes. "Rejection Sampling," one form utilized less often but still worthy of attention involves creating random data points but ultimately acceptances determined by certain requirements established beforehand are met. Generating specifically targeted outcomes has been streamlined through this relationship between randomness versus specific acceptance needs coupling. High-performing systems have emerged from its use despite its otherwise steep computational requirements revealing itself as a possible method to employ. The selection, regardless of the technique utilized, remains reliant on necessary consideration taken regarding achieving precision while mitigating potential consequences that affect both operational speed and computational resources.

Advantages and limitations of Generative AI: This section discusses the potential benefits and drawbacks of using Generative AI. It covers topics such as the ability to generate new, realistic data, the potential for creative applications, and the risks of generating biased or misleading data.

With its ability to create new data that is almost identical to its training data through deep learning algorithms, Generative AI is a fascinating field capable of revolutionizing many industries, from entertainment to healthcare. The advantages it brings along are crucial; however, we must also factor in its limitations before full-scale implementation. Throughout this section, we explore in-depth both pros and cons related to Generative AI's impact on the world of artificial intelligence (AI). We begin by evaluating benefits like realistic data generation's transformational power for creating synthetic data resembling real-world imagery at a more economical rate for various industrial sectors like images building or natural language smoothening. In contrast, on downside alert about the potential implications of biased results through compromised accuracy issues remains equally significant ground covered in this section thoughtfully crafting insights into Ethical consideration pertaining to this trend-setting development's acceptance gradually across domains affecting users in question.

Fashion designers may also benefit greatly from generative AI by using it as a tool for creating one of a kind designs at lower cost than traditional methods. Using its capability to learn from vast collections of existing clothing designs when trained on big databases, designers could generate novel design ideas quickly and efficiently, reducing overall production costs per item. To cater appropriately for clients' preferences, your team must create customizable designs for each one of them. Nonetheless, this process demands significantly more resources and time compared to regular designing methods used in creating clothing products intended for general public consumption globally; making it less cost-effective overall. Fortunately, there is an actionable solution:

97

Generative AI proves highly valuable upon enabling businesses serve their customers better by generating numerous tailored patterns based on customer preferences via programmable algorithms. This consequently leads to reduced operational expenses and faster delivery times thereby enabling wider client access with relative ease beyond initial projections manifesting as production hurdles. Development time increases alongside costs with traditional techniques utilized across various domains.

In order to mitigate delays and resource utilization concerns around asset creation effectively, generative AI shows promise as it enables faster outcomes using less input than conventional methods. With the assistance of generative models, designers can bring their vision to life in an exceedingly brief period (hours or days), which would typically necessitate years of effort spent on intricate details manually.

By deploying GANs during game development for instance, high-resolution textures or 3D models representing characters' landscapes or buildings are easily engineered within a short time frame post which further tweaks are possible such as color modifications, etc.

Similarly, in fashion designing too, dramatic progress has been made with GANs, where traditionally designers had to create physical prototypes after sketching while now GAN-generated three-dimensional clothing designs breathe life into creative aspirations with customized features duly incorporated with precision.

Limitations of Generative AI

Generative AI comes with its own set of limitations that must be considered when developing and employing these models. One key limitation is related to biased or misleading input data that compromises the accuracy of generated output.

The successful performance of generative AI models explicitly depends upon high-quality input datasets free from any bias or errors while compiling such datasets. For instance, let us take an example where a company develops a generative AI system for forecasting default loans based on historical financial figures but uses flawed historic records containing incorrect projections; thus resulting in inaccurate output predictions with drastic business impacts.

Medical enterprises use generative deep learning-based systems to analyze patient health history predicting admission rates. If any false beliefs lay in compiled patient databases pertaining to illnesses that they had never suffered before or inaccurate family medical history, it will result in high-risk output predictions for patients' health.

To evade such concerns, companies need to put additional efforts to understand, remove biases from compiled data samples, and include ethical considerations during development and implementation. Generative models face limitations when attempting to generate realistic variations from data that weren't included during their training phase. Rotated or distorted digits could demonstrate these constraints in regard to creating images most specific to digital utility (e.g., representing numbers). This limitation becomes more evident and complex for real-world applications since data diversity is higher compared to standardized sets used for testing that are usually more controlled environments (e.g., visual cat identification software made specifically for indoor cats). To enhance generational abilities beyond pre-existing data points by expanding its coverage range, researchers engage techniques such as transfer learning which use pre-trained models.

Others may go for data augmentation, which involves artificially creating new examples via transformations e.g., rotation, cropping, or adding noise. Achieving a more realistic touch in a painting using generative AI may require providing more training data, which can be time and effort intensive. In other fields like natural language generation, this AI may generate text that is inaccurate or irrelevant limiting overall control over the outcome. Therefore, while generative AI has potential and impressive results it's crucial to consider its limitations in controlling outcomes. Resource intensive is one limitation of generative AI as it needs a lot of computational power and memory to generate complex models.

For instance, a large organization may require generative AI to create realistic 3D product models for their advertisement purposes. However, training the model demands expensive hardware and software investments along with hiring data scientists and ML experts. Additionally, generating new 3D models can exhaust significant computational power leading to further resource investments. Generative AI bears ethical implications such as privacy infringement and bias especially if the training data is incomplete or biased. It's essential to examine how bias can influence individuals or groups when working with generative AI tools effectively. Moreover, facial recognition software underlines how this technology impacts privacy, by its power of generating highly accurate renderings of legitimate-looking people through its capabilities in generating new digital content. This can be used illegally and maliciously without direct consent from those featured in it as well individuals working sensitive fields like finance-related industries amongst others. Users need to remain conscious of both its exciting possibilities and ethical implications.

If one could, for example, create phony imagery using generative AI and spread it across various platforms, this could cause a lot of damage to those depicted in the pictures. It's also problematic to consider that criminals can use generative AI-created face scans to create fake identities for identity theft or other fraudulent purposes. Meanwhile, social platforms continue to harbor user-generated content and identities that serve as primary resources.

It's crucial that people in charge of developing or utilizing generative AI technology have a deep understanding of the issues surrounding usage. For instance, implementing strict guidelines around usage protocols is crucial before employing an individual's image along with consent considerations.

While generative AI might fail when it comes to producing unbiased output due to being dependent on biased training models, the major issue it poses concerns its limitations regarding tracing such biases when they come up impacting society heavily. This leads toward perpetuation throughout several fields like healthcare practices, criminal justice systems. Regrettably, despite serving a vast demographic, this website has failed to incorporate critical accessibility provisions catered toward persons with disabilities. Design flaws such as these overlook an integral part of its user base and oppose the basic value proposition that requires equal consideration to be given to all users: inclusivity.

The potential impact of Generative AI is immense across various industries. However, just like any other technology that shapes society, Generative-AI equally carries ethical considerations too. Bias can be a detriment within this evolving technology resulting from erroneous data. The output results from these algorithms rely greatly on accuracy and unbiased inputs eventually affecting crucial decisions within certain industries. Consider a generative model developed specifically for hiring practices; it might just select male candidates for a job simply because it may have been trained primarily with datasets containing resumes of male candidates rather than focusing completely on their competency.

Similarly, Facial recognition technology evolved as an advantageous tool for law enforcement officials but has since raised alarm bells due to unjust arrests made after being fed into biased systems without enough diversity concerning ethnicity representation within its source data sets, which makes it unreliable overall.

The importance of accurately represented datasets cannot be overstated as one vital way to mitigate any sort of bias trickle within Generative-AI models. Trained models need proper testing along with active updation to counteract any biases that might creep in. Additionally, human intervention remains pertinent for ensuring the avoidance of perpetuating gender or societal stereotypes. Generative AI has the potential to bring creativity but greater discretion is key to avoiding it being used as a tool for fake news or propaganda. Let us consider an era where anybody with access to Artificial Intelligence (AI) tools can effortlessly craft seemingly realistic news articles merging seamlessly into non-fiction. This consequently leads us down a dystopian path where journalism and media lose much of their credibility as mischievous elements exploit this new power by disseminating malicious disinformation. This significant shift of power could effectively facilitate considerable public opinion manipulation and make high-stakes political elections vulnerable to foul play such as deepfake technology. This technology is capable of fabricating videos featuring individuals saying things they never did or said, thereby harming their reputation in society. Chatbots pose another significant threat since orchestrated maliciously; they could produce false conversations and comments on social media channeled toward amplifying a particular viewpoint while simultaneously discrediting opposing views.

It is essential to first understand the capabilities and potential pitfalls behind generative AI technology to pre-empt misuse that might arise in later application. Things such as ethical guidelines and regulations must be defined to forestall its use in spreading falsified information, destroying reputations or promoting unethical behavior that could ultimately destabilize society at large. Researchers, industry leaders, policymakers alongside the general public must collaboratively foster responsible AI practices through all stages of development.

Particularly concerning is the ownership rights regarding copyright infringement issues arising with generative AI models' application. For example, using an unlicensed model trained on copyrighted data may result in an artist using it to generate an exceptional painting that takes off on social media and consequently generates substantial profit. The original creators of data used may deem it appropriate to seek compensation for their efforts from any profits made unlawfully through unauthorized access. Suppose that a corporation utilizes artificial intelligence (AI) technology for creating musical pieces. One of its attempts with generative AI systems generates an instant megahit song that receives many accolades from critics as well as fans alike. The inquiry now arises: who possesses the right over this tune? Does it belong solely to the company or does some of these rights belong to the AI model as well?

Consequently, resolving questions about ownership and copyrights necessitates meticulous deliberation along with following proper regulations that may prevent legal conflicts from emerging in such situations—especially when dealing with generative AI technology.

Furthermore, ensuring transparency regarding data utilized in training models could prove crucial toward retaining ethical behavior when producing digital content like music, art, or other areas dependent on generative AI tools.

Nonetheless, despite some prevalent challenges arising through implementing it, Generative Artificial Intelligence systems can significantly transform several industries from Design to Medicine for such creative purposes. This is, if it is established by considering potential advantages alongside setbacks critically while being realistically conscious about acting ethically-responsible throughout all aspects of usage alike.

Applications of Generative AI: This section provides examples of real-world applications of Generative AI, such as image and video synthesis, natural language generation, and music composition. It also discusses the potential for Generative AI to be used in fields such as medicine, engineering, and art.

The growing use of Generative AI has not only extended beyond just one sector but also across numerous industries as well as varying fields. In this section, we will explore the most innovative uses of Generative AI witnessed so far throughout our journey including image and video synthesis—a rapidly evolving area with much excitement. This refers to the use of machine learning models in creating new images or videos that appear useful and appealing. Regarding image synthesis, we have witnessed impressive results from StyleGAN, a model which creates high-quality photorealistic images of human faces that do not exist initially.

StyleGAN uses data set patterns learned from actual human faces enabling it to generate even more diverse ones. This technological advancement equips artists and designers for developing authentic graphics without breaking the bank on licensing fees and photo shoots. Conversely, video synthesis is much more complicated as it requires beauty in images that are compatible with observable changes over time. Examples include DeepDream which uses deep neural networks in generating surreal videos while DALL-E model incorporates natural language descriptions.

This generative technology is useful beyond the artistic dimension and has successfully permeated several other fields including entertainment, advertising for one sake, and possibly medicine as well where realistic images of a patient's internal organs can aid in diagnosis and treatment options by doctors.

Nevertheless deep concerns arise around improper use of these technologies resulting in deepfakes. They are synthetic videos that seem real but feature manipulated footage created from real existing persons raising potential risks over misinformation and other harms caused through deception highlighting the responsibility each creator must possess.

Natural Language Generation (NLG) also emerges as an essential branch under Natural Language Processing (NLP) tasked toward teaching machines a method to reproduce text with apparent human-like style placed intact. The numerous applications offered by natural language generation (NLG) make it an exceptional tool across different industries.

Advantages of NLG:

1] This system can create custom messages via chatbots or personal assistants along with the automatic creation of reports or articles.

2] One practical example is utilizing these systems in designing tailored marketing emails for businesses dealing in e-commerce platforms

3] This method involves using available consumer data recorded by their various correspondences while shopping on the website.

It is a potential means to promote products via messaging fused with pertinent promotions suitable to what the customers explicitly prefer based on their recent transactions giving them a pleasant customer experience.

4] Journalism has immensely benefited from automated technology and NLG is playing a prominent role in automatic journalism technologies currently

5] For example, these systems can generate detailed and accurate sports articles that involve statistics, highlights, and scores within seconds. This can assist news agencies in publishing quick news.

6] In tandem with healthcare settings, NLG systems permit practitioners to summarize vital information about patient health conditions immediately. It's significantly helpful for the medical staff when they need to review records immediately without taking up much of their time.

Challenges:

However, NLG faces its own challenges as it labors with understanding complex language structures like sarcasm or idioms that hinder the output accuracy leading to confusion or misinterpretation of text in some cases. Also the ethical implications relating to unbalanced selection contained in the training data could affect future decision making by AI models.

With Amper Music's intuitive platform powered by an advanced AI algorithm for music generation, users can input different parameters such as mood, tempo, or genre not only to access custom-created pieces but also use them effectively across commercials or media projects with relative ease.

The application of MuseNet generated via OpenAI has expanded possibilities for us that is beyond comprehension concerning the style and genres it covers such as classical music down to Jazz or pop genres respectively. This diverse deep neural network modifies musical compositions inspired by given melody outlines based on its previous learnings from training sets containing sheet notes alongside MIDI files.

The impact generative artificial intelligence makes in creating novel compositions heightens through abilities like drug discovery/development optimization. Its power to create synthetic medical images without exposing patients to potentially harmful radiation or times when invasive surgical procedures would ordinarily be required without the option using various large dataset medical images to generate accurate and realistic imagery in support of diagnosis and therapeutic intervention is notable. Researchers today have an innovative tool at their disposal: generative models that allow them to create new molecules for testing with therapeutic potential that could offer hope for better outcomes than those currently available. This technology also has broad application in personalized medicine by providing doctors with essential insight into factors relevant to disease susceptibility markers. This leads to tailored therapy plans aimed at improving recovery prospects while minimizing harm from adverse effects.

Engineering applications have also seen significant benefits from their use of generative AI-driven virtual-models instead of relying on time-intensive simulation methods previously used as part of this iterative process. This results in products designed more efficiently thanks again due mostly to reduced costs during the development phase as well.

Machine learning algorithms are powering the future across industries today.: A particular case-in-point relevant here is "generative design" leveraging these same technologies effectively transforming large amounts of data into an incredibly vast number of design options. This methodology has applications in automobile manufacturing as it provides designers with many Aerodynamic body designs, safe and sustainable, thanks yet again to varying data points being analyzed correctly.

We're yet to see where Generative AI's limits lie! One example is how organizations can optimize their present systems with this technology's assistance; such possibilities result from analyzing vast quantities of data via machine learning algorithms directed toward system behavior investigation.

The employment of modern technologies within art creation sharpens artists' ingenuity better than usual boundaries imposed by traditional methods; generating new compositions by using Generative Adversarial Networks (GANs) is an excellent option here. Utilizing two hybridized neural networks—one providing content and the other assessing it according to authenticity—modelers can generate accurate outputs based on trained datasets. An example of this concept's potential is the artwork created by Robbie Barrat, where he utilized a GAN trained on classical portraiture, leading to unique artwork.

Recently, Generative AI has also given rise to novel pathways within music creation without relying on human direction. By feeding music samples into created neural networks, one could train them in pattern recognition and structure learning; thus, it becomes possible to generate new compositions that are fresh as well as innovative entirely! Holly Herndon's album "PROTO" is an excellent example of AI-driven creativity influenced by multiple researchers and utilizing human vocal inputs.

Digital animation and Digital art are both possibilities for further development of Generative AI's potential impact; there will undoubtedly be more applications ahead for this technology! Advancement in technology has allowed artists to create highly-detailed animations with ease through deep learning algorithms—something previously prohibitive without them. Notably Sofia Crespo's 'Neural Zoo' project, which features a collection of imaginary exotic animals simulated by a deep learning algorithm trained on marine wildlife images exhibits this capability perfectly well.

While numerous prospective applications exist for Generative AI across several domains from music composition to generating natural language synthesis among others, caution should also be warranted considering associated downsides and ethical challenges.

For instance, data privacy violation as well as biased information output may arise from using Generative models within healthcare settings while inappropriate deployment on content creation stages could infringe upon existing copyright laws or other ownership considerations too.

Generative AI technology presents tremendous transformational possibilities. Promising advancements in numerous fields like education and medicine amongst others can be realized by utilizing this technology. To achieve truly groundbreaking and ethical results, careful deliberations are necessary amidst the potential inherent risks for every application and stage of deployment.

Future developments in Generative AI: This section looks at some of the emerging trends and developments in Generative AI, such as unsupervised learning, meta-learning, and transfer learning. It also discusses the potential for Generative AI to continue advancing and becoming more powerful in the future.

Generative AI has made significant strides since its inception and impacted numerous fields such as image and speech recognition, natural language processing and drug discovery positively. The continued progress in technology and access to an immense amount of data suggests that Generative AI will experience considerable growth in the years ahead. Here, we examine some emerging trends reshaping the future of Generative AI.

Unsupervised learning has arisen as an essential trend within Generative AI that aims to address inherent limitations in current models' reliance on labeled training datasets. By training on unlabeled input samples instead, unsupervised learning enables machine learning models to recognize patterns automatically without explicit instruction; thereby allowing them to identify subtle and nuanced patterns eventually. Contrastive Predictive Coding (CPC) is one method showing great potential as part of unsupervised learning within Generative AI. It emphasizes developing meaningful autonomous representations of diverse sets of data by predicting future latent representations from earlier ones in a given sequence context.

CPC utilizes a contrastive loss function alongside negative samples aimed at encouraging model feature development toward greater discriminant features: leading ultimately toward better generalization capabilities when encountering newer or previously unseen types of data. CPC is a promising tool that could find new applications in speech recognition and natural language processing. By implementing CPC to learn speech and text data representations, it could lead to improved accuracy and efficiency in these fields.

Meta learning is a rapidly evolving area of research in generative AI that has the potential to revolutionize how machine learning models are trained. Essentially, meta learning works by teaching an AI model how to learn better. The model does this through learning new skills faster by building on its previous knowledge and experience, which leads to better performance overall.

Picture yourself trying to teach a child how to ride a bike. At first, they may struggle with balancing, steering and pedaling all at once but with practice comes improvement. When presented with a new bike that has different features like handlebars or wheels, they may still be able to adapt more quickly because of their learned core skills. Our goal with meta learning is the same: we want our AI models not just to be proficient at one specific task but also to excel at multiple tasks because they've learned fundamental skills that can be applied across various situations. One example of this technique is being used for few short learning in generative AI where data sets can provide limited information about particular subjects or skills.

In another instance, imagine training an AI model on recognizing dog breeds; however, it struggles when introduced to breeds it hasn't seen before. Through using meta learning techniques though, the model will begin adopting past experiences it had with other breeds and apply them toward recognizing the new breed as well.

Finally, another area where meta learning techniques can be applied is reinforcement learning, which aims at training decision making based on environmental feedback. This allows fallible AI models to make necessary adjustments for optimal performance parameters relevant throughout applications today! Efficiently learning from the environment by using meta learning is an essential aspect for machine learning models to adapt quickly to new situations.

Transfer learning is a technique in machine learning where we reuse an already trained model on a related task instead of starting over each time from scratch. Generative AI uses transfer learning efficiently by leveraging knowledge learned from prior tasks during training, allowing us to train models with less data efficiently. For example, if we want our AI system to generate lifelike images of dogs first, we train it on breed classification using a vast dataset containing labeled dog images before training it for generating new dog images with less labeled data examples.

The prior knowledge gained during this breed classification stage is transferred as part of the subsequent generative tasks input making it learn faster while requiring fewer labeled examples only. In addition, transfer learning can be used across multiple domains or tasks; any pre-trained Dog image generation model could be used as starting base parameters in a Cat image generating task. Transfer Learning technique potentially increases efficiency and helps improve performance and generalization for later use cases effortlessly.

Generative AI has benefitted hugely from transfer learning techniques development over time; one perfect example being Pretrained Language Models (PLMs) such as OpenAIs GPT 3, which excels at generating human-like text articles. PLMs rely on an essential idea—they get trained on copious amounts of textual material, encompassing books and websites alike. As this training progresses, the model grasps text patterns, including word relationships and nuances behind certain expressions, ultimately developing an extensive knowledge set about language that can serve multiple objectives.

One factor that makes them highly advantageous is their versatility as they're trained using diverse inputs; there's greater latitude when it comes to producing content across different domains, for instance, news reports or creative writing etc. This is why enterprises and other establishments find it very handy while handling sizable volumes yet ensuring quality output. In regard to future developments within Generative AI's landscape—noting how much impact they already wield—their significance will be amplified even further by initiatives regarding thorough improvements like refining algorithmic protocols and integrating variegated datasets. This facilitates accurate predictive outcomes alongside being intermingled with diverse applications ranging from chatbots to virtual assistants. These factors as a whole makes them an integral part not just for SMEs but all types of businesses.

Generative AI for Scientific Discovery

Generative AI also demonstrates potential when designing new materials or molecules without the usual trial-and-error approach commonly carried out at significant cost and effort invested over long durations. This method paves new avenues for cost-effective advanced material development pushed forward by machine intelligence. Generative AI contributes to accelerating the discovery process for novel molecules and materials, through predicting their behavior and properties even before their actual synthesis takes place. One crucial approach that aids researchers in this regard is known as Deep Generative Models. These models learn from vast datasets containing information about existing compounds along with their predictable behaviors/properties.

Such models then bring into existence fresh compounds displaying predetermined desirable traits that otherwise would've taken quite some time to find. Researchers repetitively execute such a generation followed by testing trails until they achieve an optimized compound with the desired requirements which can then lead to its synthesis/production/commercialization. University of North Carolina researchers leveraged Deep Generative Models in predicting organic molecule structure tremendously improving its performance while retaining functionality crucial for solar energy effectiveness.

An interesting way scientists are predicting protein structure involves using generative AI models such as AlphaFold by DeepMind. This model employs a deep neural net to predict their three-dimensional composition from their amino acid sequence after being trained on available protein structure databases creating avenues toward faster drug discovery treatment development.

Computers can also be engineered for other related applications like molecule design generation when researchers simulate an atom's behavior allowing quick production rates leading toward creating previously impossible material types.

Generative AI can be inspiring in fields related to creativity such as fashion, art, and architectural design when it comes to generating only original ideas. Designers extensively use generative AI models in producing eye-catching and unique clothing that is different from the norm and that can capture the world's attention within a few seconds.

By predicting fashion trends through generative AI, designers can create impressive marketable designs ahead of their competitors through analyzing historical data and identifying patterns from consumer preferences ensuring they keep up with current styles while retaining market dominance.

One of the most fascinating advancements of Artificial Intelligence is Generation. In the fashion industry, designers are leveraging Generative AI to develop new concepts while exploring design spaces not previously considered. This groundbreaking technology allows creatives to push boundaries further ultimately leading to pioneering clothing designs that captivate public imagination.

Additionally, by automating specific aspects of fabric selection and pattern making using Generative AI, designers optimize their manufacturing processes resulting in a more efficient DMP without compromising on product quality hence contributing toward a bottom line boost in profitability.

Another advantage is personalizing clothing tailored toward an individual's preference through analyzing data such as tastes and styles via Generative algorithms leading to individualized products and customer satisfaction, which is becoming essential nowadays. The applications available through Generative AI are endless prompting widespread use across various industries today; there may be some concerns arising from utilizing Gen AI given its exponential growth rate for this technology. It's therefore advisable that stakeholders be vigilant i.e. cautious with regards to ethical considerations while remaining proactive about keeping up with latest developments and trends concerning transparency on the minimization of these issues/conflicts.

Ultimately, we must be aware of potential disruptive impacts so as to utilize this technology optimally while minimizing uncertainties along the way.

Case studies of Generative AI in various industries: This section provides examples of how Generative AI is being used in different industries, such as advertising, entertainment, and finance. It showcases the impact that Generative AI is having

on these industries and the potential for further growth and development.

Generative artificial intelligence (AI) has brought forth noteworthy advancements across industries—finance operations streamlined and fashion aesthetics improved. This technological innovation helps organizations explore innovative pathways when making strategic decisions; its unquestionable potential are illuminated through many successful case studies detailed below.

- One sector that consistently searches for new ways of engaging consumers through unique visuals is advertising; hence, Generative AI has emerged to be a powerful tool for the industry. A notable success story is H&M clothing brand utilizing Generative AI to compose original scores for promotional videos. This technology scrutinizes the hues, patterns, and shapes within the video to create customized musical tracks that complementarily blend with visuals—resulting in more memorable advertisements elevating H&M's unique standing among its competitors.
- Similarly, L'Oreal introduced "Perso"—a personalization tool creating tailor-made skincare products adjusting depending on skin type, customer preferences and illnesses through Generative AI—a novel avenue that offers an enhanced experience for customers while allowing L'Oreal to differentiate itself within saturated market competition.
- The digital-advertising realm has also missed no opportunity to integrate generative AI technologies into optimizing personalized advertisements as companies such as Adgorithms and Persado use data analysis tools in their practices. Subsequently, these novel approaches lead to higher engagement rates; thus providing favorable return on investment opportunities.

- The entertainment industry is inherently founded upon creativity and spectacular delights; however with the pioneering efforts of generative artificial intelligence, technological innovation only opens up new avenues of creativity and facilitates unmatched possibilities unparalleled previously.

- Generative AI is transforming finance in numerous ways; from detecting fraud activities through to expertly optimizing portfolios. But one area attracting particular attention surrounds its ability to predict stock price changes, which traditional methods have struggled to do reliably using only past payment history data alone. By incorporating datasets from social media feeds along with other publicly available information sources, it's now possible to gather time sensitive market insights, which allows those investing in these areas an increased chance at maximizing their returns.

- An area worth noting surrounds credit risk superintendence, another field once almost entirely reliant upon historical data, e.g., credit scores or payment histories. Yet now, since they have access to wider-ranging analysis through additional datasets (including online behavior) from Generative AI algorithms, they're able to offer even greater insights into individual creditworthiness levels.

- The broad array of benefits offered by Generative AI has been noted across several sectors including finance, where it serves multiple purposes such as investment portfolio optimization utilizing sophisticated analytics software designed around balancing ROI with a reduced risk profile among others.

- Generative IV's contribution isn't just restricted within financial entities but extended applicability exists across diverse segments like customer-service enrichment; fraud prevention as well as adherence to regulatory standards.

- Its influence cuts across several business segments and industries, contributing a new dimension of creativity in decision-making and accelerating innovation.

Generative AI's sweeping potential applications herald an innovative era with endless possibilities. Nevertheless, it is important to sustain responsible usage while factoring ethical considerations as artificial intelligence continues to present itself as a significant facet of contemporary society.

REINFORCEMENT LEARNING

An Introduction

Reinforcement learning (RL), recognized among specialist circles for decades now has recently exploded in popularity concerning machine-learning techniques; this revolutionary approach teaches agents "learn-by-doing" via trial-and-error environments where agents receive feedback-punishments/rewards via specific interactions. The end goal is for these agents ultimately selecting optimal actions that lead up to the highest possible cumulative reward achievements.

In this section, our aim is to introduce you to agent designs—interactions between environments-and-precision reasons why rapid growth occurred ensuring widespread adoption in recent years within gaming, finance, and robotics, among others. Meanwhile, we delve into critical components of reinforcement learning through shared insights along with associated algorithms and related challenges in developing RL agents.

The three pillars of reinforcement learning: (1) the agent (2) the environment (3) rewards—from its foundation; this includes the exchange of information necessary for an agent to learn while interacting with a given environment.

The driver behind RL lies in an agent taking action while receiving rewards/punishments as feedback—this insight helps guide agent actions toward optimal cumulative reward accumulation steadily over time. The process consists mainly of five phases: 1-observation; 2-action; 3-reward; 4-update; and 5-repeat till end-episode or task completion.

Several algorithms exist under the RL umbrella; selection depends on strengths and weaknesses in specific situations where picked optimization is vital. We cover different types to give you a comprehensive view. If you're seeking more insight into reinforcement learning (RL), it is important to understand its algorithms used widely—one such approach being Value-Based RL, which focuses on training agents in decision-making situations. This method involves having an agent learn a value-function capable of estimating potential long-term rewards stemming from every state within an environment. This ultimately results from actions taken by agents that maximize learned-values leading it toward forming policies with maximum reward levels.

Let's tackle a scenario through Pac-Man gameplay: At the start of the game session, pacman randomly begins on the game board cluttered with pellets, power-ups, walls and ghosts. The objective is to eat as many pellets while avoiding the ghosts and walls to max out its score.

To train an agent for this task, we endow pacman with a reward function awarding positive scores for each ingested pellet and negative score for ghost hits; all coupled with additional penalties per time-step. The feedback signal garnered from the environment aids our pacman agent in updating the value function where associated values reflect future reward expectations resulting from any given state.

Using a neural network representation of this value function, crystalizes Q-learning. This is a technique that trains agents across action selection by optimizing expected values garnered through received rewards while also examining estimated upcoming states' worth values. Iteratively optimizing during training sessions leads to improved accuracy across decision-making possibilities resulting in our agent achieving high scores based on optimal policies obtained through Value-Based RL's ability to learn continuously from experiences.

Value-Based RL has proven its ability in handling complex decision-making tasks necessary across real-world applications such as finance or autonomous driving by allowing continuous learning from experiences gathered over time. When applying Reinforcement Learning (RL), one approach employed is Policy-Based RL, which requires an agent that learns how best to make decisions through interaction with its environment. In executing this strategy, agents establish "policies" which map out possible actions based on encountered states with specific priorities given to certain decisions when aimed at maximizing cumulative rewards received from their environments over time.

Let's say we're playing chess—with Policy-Based RL implementation—the computer will represent our "agent" whose interactions take place on the on-board interface serving as our "environment" with each piece having varying states dependent on their respective positions. The movie chosen by our "agent" serves as its "action" at every point in the game with the final outcome— win, loss, or draw—serving as our ultimate reward for playing through the iteration.

Initially, our "agent" starts from a random policy before trial and error activities lead to updates of these policies after feedback is obtained. Successful "actions" earn increased weight, increasing the likelihood that they'll be selected in future while "actions" that fail get weighed down and receive less consideration when encountered again.

The REINFORCE algorithm is often used within policy-based reinforcement learning else, it can handle continuous action spaces when not easily described as value functions. This is an advantage in complex environments where it can be difficult to predict optimal policies. This technology boasts versatile applications spanning numerous fields, such as robotics, finance, and game playing.

Actor Critic Reinforcement Learning (RL) is a powerful algorithm that combines the advantages of both value based and policy based methods. The approach involves two important components: the actor, which selects actions through policies and the critic, which evaluates the actions' value. Consider playing a maze game where reaching a goal is the objective. The actor mirrors the player's role in making moves in pursuit of this goal. Meanwhile the critic represents a coach who monitors these actions and provides feedback on their effectiveness.

In this RL, algorithm policies direct how to navigate through the maze while evaluations from critics gauge progress toward achieving goals. As time progresses, players learn from feedback received from their coach and adjust decision making for optimal results. Within machine learning contexts, Actor Critic RL is ideal for training AI agents that play chess. Model Based RL involves an agent's effort to learn about environment matters so as to decide on maximizing reward signals. This method sees agents predicting action consequences before choosing which action to execute. However, it encounters challenges if not applied prudently. For instance: having robots learn to navigate mazes requires careful measures when applying Model Based RL since unwise decisions affect target achievement significantly.

The robot follows a set course designed to reach its goal quickly with obstacle-laden pathways that require navigation skills at each checkpoint along its journey. These skills include differentiating forward motion from precise left or right turns at critical sections of the maze.

Model-Based Reinforcement Learning (RL) techniques come into play using a data collection strategy from exploration combined with actionable options within each sequence. From this dataset collection experience, arises the creation of comprehensive robotic models with pinpoint accuracy detailing maze states, while incorporating unique transition points between each state.

For agents seeking optimum outcomes, this approach proves more efficient because they make informed decisions based on prior collected data experiences leading to specific results. This is very useful for improving decision-making processes across industries worldwide for agents seeking operational success across a range of domains requiring varying execution needs such as timely resource allocation or task executions while prioritizing safety concerns!

Notably, challenges exist in building accurate robotic models under ever-changing dynamic environments such as altering rules over time. Scholars and researchers worldwide are actively exploring ways to improve model accuracy and flexibility by utilizing deep learning models offering optimized inputs capable of automating repetitive tasks while providing cost-effective solutions across sectors and industries globally!

Challenges in Reinforcement Learning

Reinforcement Learning is not without its challenges, one of which is the Exploration-Exploitation trade-off. Imagine that you're playing a game as an agent; you can either choose known actions with defined rewards or opt for exploring new ways for higher returns. Not exploiting current knowledge, promotes suboptimal performance while constant exploration may mean missing out on the potential rewards of learned experiences. The Epsilon-Greedy algorithm manages this situation by reducing exploration probability over time as experience increases. Researchers continually develop new algorithms and approaches toward effective learning in complex environments.

Credit Assignment also poses a critical challenge where successes rely on delayed reward signals from an action taken by the agent. Credit assignment is a crucial aspect of developing effective reinforcement learning (RL) systems because agents need credit for taking actions that contribute most toward desirable outcomes. Take, for instance, robots operating under various scenarios that require turning left or right when navigating mazes of different lengths. Assigning credits based on delayed feedback signals present challenges since the requirement is that we carefully consider several possible factors that might influence the outcome.

Temporal difference (TD) learning and eligibility traces can help solve this challenge by taking into account an agent's entire history of actions while accurately crediting each action's contribution toward achieving desirable outcomes.

In addition to credit assignment, generalization poses another critical challenge in developing robust RL systems—ensuring agents can apply their learned policies correctly in new situations beyond training data. Chess provides an excellent case study for illustrating this point since RL agents must learn generalizable strategies beyond mastering standard opening moves by applying these tactics effectively against unfamiliar opponents with distinctive playing styles. To ensure that an agent is effective in playing against multiple opponents or in different environments, it must possess the ability to generalize its strategy.

One way to achieve this is through transfer learning where the agent is trained on multiple related tasks or environments allowing it to develop a more generalized policy. Alternatively, function approximation methods such as deep neural networks can also aid in generalization. However, generalization remains a challenging problem in RL and ongoing research seeks to improve agents' ability to effectively generate solutions for a wide range of environments. In Reinforcement Learning (RL), dealing with partial observability presents one of the most significant challenges for researchers and practitioners.

This limitation occurs when agents cannot observe the complete state of their environment, reducing their effectiveness in making optimal decisions. For example, when playing poker or driving self-driving cars, partial observability creates difficulties that must be overcome using POMDP techniques or deep neural networks that enable learning representations that are robust even with limited observations. By processing raw sensor data and acquiring appropriate environmental representations, agents can achieve more successful decision-making outcomes in cases where they lack full observation capabilities. When deployed within robotics or other applications where cameras are used for observation purposes, limited field-of-view remains a concern that impacts overall agent effectiveness. Nevertheless, thanks to deep neural networks and their capacity to capture and assimilate these critical environmental details from captured footage—agents can take into account this vital input and inform their choices accordingly.

The History and Evolution of Reinforcement Learning

Across several decades, the story of reinforcement learning plays out richly with fascinating paradigms stemming from research around adaptive control systems and optimal control theory that began back in the 1950s. These early works aimed at developing automatic adjustment schemes that improved performance over time—essential stepping stones leading toward creating what we now know as RL today.

Machine Learning (ML)'s birth sparked curiosity among researchers interested in developing intelligent agents capable of learning and adapting to their surroundings during the need-to-know era of AI studies in the 1980s. Amongst them were those like Christopher Watkins who were interested in exploring how ML could be applied toward solving RL problems – and whose impactful contribution came with his introduction of Q-Learning back in 1989. This model-free, off-policy RL algorithm's simplicity remains an invaluable asset today as it uses state-action values as a guideline to deriving optimal policies for agents.

The fervent activity of the 1990s was marked by the introduction of neural networks as function approximators in RL algorithms for complex value functions. Researchers were quick to accept this valuable new tool, which presented promising possibilities across a wide range of applications such as robotics. Here, researchers began exploring RL algorithms' practical application in real-world robots to help them learn complicated behaviors adaptively under different environmental conditions.

Cutting edge techniques like actor-critic methods that combine policy-based and value-based techniques yielded superior results around the last decade—emphasizing research around developing new algorithms and their applications. A broad range of industries—from robotics to autonomous driving—have harnessed actor-critic methods for successful Reinforcement Learning (RL) implementation today. The growth potential is huge, considering how interdisciplinary this field is: encompassing computer science, statistics, neuroscience and psychology to name a few. Healthcare and finance also benefit from RL applications.

The origin story goes back about 100 years when 'behaviorism' became the primary school of thought in psychology—emphasizing environmental influences on observable human behaviors with some significant implications on RL design principles. Behaviorist Tenets reinforced experimental approaches that helped agents learn from their surroundings through error correction mechanisms empowered by reinforcements (rewards or punishments).

Operant conditioning provides an example to highlight this notion where agents adjust tactics based on feedback received from the environment. This shows similarities with trial-and-error-learning experiences reminiscent of the definition espoused by Behaviorism since these adjustments occur based on environmental feedback. Although Behaviorism shaped RL algorithms such as Q-learning, it failed to consider cognitive processes that impact human thought and action performance. In recent years, professionals interested in artificial intelligence have been turning toward modern examples from cognitive psychology and neuroscience to comprehend how people make decisions—particularly through reinforcement learning techniques. During the middle of the 20th century, research by prominent American psychologist B.F. Skinner introduced principles of operant conditioning, which found that behavior depends critically upon contingent rewards paired with deficient punishments following undesired behavior. By using rewards or punishments, it was found that animals could be trained for specific tasks like pressing a lever for food.

It was also determined that the effectiveness of this technique (known as operant conditioning) relied on the timing and consistency of these rewards or punishments when shaping animal behaviors. Its applications span across various fields including education, business, and healthcare. Within educational settings, positive reinforcement can influence student engagement on a task whilst negative punishment can deter disruptive behaviors whereas within companies, it boasts of management motivation toward employee performance outcomes through reward or punishment schemes. Therapists frequently adopt operant techniques within clinical settings too—manipulating habits with reinforcements where patients need help adopting healthier behavioral patterns—such as drug addiction recovery programs etc.

Previously, the mid-20th century disputed Behaviorism for its inability to elucidate complex human behaviors; however motivation via reinforcers remains an effective way organisms alter their behavior toward their environment.

The Birth of Reinforcement Learning:
The Multi-Armed Bandit Problem

First introduced in Richard Sutton's and Andrew Barto's book, "Reinforcement Learning: An Introduction," reinforcement learning refers to a field whose origins date back to the introduction of the "multi armed bandit" problem during the mid-twentieth century. Classic applications require gamblers selecting from various slot machines each with different payout rates while trying to maximize their winnings through exploration of new machines and exploitation of known high yield ones.

Similarly, investors must handle investment opportunities compared to successful investments while advertisers can face comparable complexities when choosing between profitable or innovative ad strategies. To address these problems, reinforcement learning algorithms encompass everything from simple greedy tactics through sophisticated probabilistic approaches. Furthermore, recent years have seen increased attention focused on reinforcement learning given its relevance within fields such as control theory, robotics and artificial intelligence. Temporal Difference (TD) Learning is an exciting development in modern reinforcement learning that updates value functions based on the discrepancy between predicted and actual rewards.

Consider yourself trying out new food at restaurants in an unfamiliar city without any prior recommendation from anyone close. After visiting various food joints, analyzing your experience along with your likes/dislikes help form certain patterns that could guide further decisions around eating places. An analogy can be drawn from this situation toward TD Learning in reinforcement settings.

TD Learning proves useful when utilized as a training mechanism for machine agents engaged in reinforcing environments like gaming arenas. The AI agent receives prompt feedback regarding each action taken while playing games enabling it for consistent improvements derived entirely from past experiences by altering estimations of future state values based on received scores. To optimize decision-making during TD learning within reinforcement systems, updating estimated values based on observed rewards ensures that actions aligned with higher reward probabilities are prioritized over others. Additionally, Q-learning operates via a model-free approach wherein it learns the optimal action-value function through trial and error.

It has gained traction as an effective solution for complex problems across different domains due to its ability to learn the Q-values for state-action pairs in an environment. Q-learning has found widespread application in various fields such as creating game-playing agents, controlling robots, and designing autonomous vehicles. Its capability of gaining new insights from experiences and adapting responses based on long-term rewards make it an efficient option for dealing with several reinforcement learning problems.

Foundations of Reinforcement Learning: Markov Decision Processes

Imagine that you're a visitor, and this is your first time in a new city. You have a map handy, but you're unsure about where to go and how to get there most efficiently. Your goal is to experience as many exciting places as possible during your visit, but you face obstacles such as traffic or crowds that may deter your progress. In this sense, navigating through the city can be compared to an agent working toward resolving problems in reinforced learning by achieving their objectives within environmental limitations.

Markov decision processes (MDP) offer an excellent model for formalizing decision-making for such circumstances with four central components-i.e., states, actions, rewards, and probabilities. States depict detailed environmental situations one might encounter while meandering throughout a given territory, whereas actions signify transit options available for relocation between two states considered on any given moment.

Concurrently, rewards reflect outcomes of agents' choices driven at different stages of sequential-states; however potential gains aren't guaranteed every time due to unpredictable hindrances like unforeseen traffic blocks resulting in time loss etc. In other words, even after selecting optimal locations based on personal preferences—enjoyment levels might still decrease further upon reaching destinations depending on various factors like traffic disruptions that could ultimately lower achieving the intended purpose sought after visiting that location.

In summary—MDPs enable agents with decision-making strategies by influencing future outcomes cumulatively with expected returns driving their actions by defining opportunities at each state-transition level concerning balancing immediate short-term profit motives against more critical long-term goals.

The process of solving MDPs requires developing policies directing the agents' behavior based upon estimated future-cumulative gains within environmental limits using mapping techniques that allocate respective choices/actions appropriate upon varied situations encountered i.e., while touring diverse locations within the given territory/region visited. To master reinforcement learning algorithms like an expert, one should be familiar with Markov Decision Processes (MDPs). This widely used technique enables an intelligent agent to optimize its own decision-making process essentially by trial-and-error approach! By exploring its surroundings via action-taking and reward-receiving feedback models, the agent can boost performance over time by modifying its policy or value function after every operation or inquiry outcome yields data-driven insights. Thus, this improves decision making skills resulting in optimum rewards gained vs loss incurred as consequences.

The Exploration-Exploitation Dilemma in Reinforcement Learning

Reinforcement learning involves training agents on environmental interaction via repeated experiments. Agents' primary aim should always be attaining maximum cumulative reward while continuously exploring their surroundings seeking crucial information. However, some decisions pose a conflict between exploiting what we know versus exploring alternatives—commonly termed "the exploration-exploitation dilemma." For instance, locating fine dining locations within unfamiliar neighborhoods involves either picking restaurants randomly or consulting locals' opinions, but both actions need to be analyzed before making a final choice. Embedding the balance between these opposites is vital and often formalized as exploitation and exploration in reinforcement learning.

The epsilon-greedy algorithm is one popular method that selects an action with a higher anticipated reward using well-defined criteria but explores less frequently at random moments. The UCB algorithm uses past experiences to compute each action's upper confidence bound and balances its exploration-exploitation trade-off by opting for actions with more uncertainty. However, maintaining this balance can be complicated, more so in complicated environments.

Over-exploration may waste time following sub-optimal options, whereas over-exploiting discovered benefits could lead to missed opportunities of stumbling on better alternatives. The best way for chess players and medical practitioners alike to succeed is by striking a delicate balance between expanding their horizons with new approaches without abandoning rock solid theories entirely. On one hand, rookie chess players may try out different strategies without much reflection while seasoned ones know beneficial moves and techniques in detail yet stay receptive to novel insights.

Likewise, healthcare professionals must decide between introducing untested procedures into their practices alongside more standardized ones that constitute established protocols. Nonetheless, you need to bear in mind that any new approach has its inherent risks as well, so every decision should be made after careful deliberation.

Dynamic Programming: Solving Reinforcement Learning Problems Optimally

The agent's objective is to maximize its cumulative reward over time through a trial and error approach. This methodology has proven successful in many fields, including gaming, finance, and robotics. Dynamic programming is an efficient means of optimally resolving reinforcement learning issues. By dividing complex optimization problems into smaller subproblems, dynamic programming algorithms assist with this process. Dynamic programming methods are frequently employed to determine optimal policies that enable the agent to make the best possible decisions in each environment state. Broadly speaking, there are two types of dynamic programming techniques: value iteration and policy iteration. Value iteration calculates the optimal value function in an iterative fashion that represents an expected cumulative reward from every state. Policy iteration works differently by alternating between policy evaluation and improvement stages to identify the optimal policy achieving maximum expected cumulative rewards.

Dynamic programming techniques have been useful for a broad range of reinforcement learning applications: game playing, autonomous navigation, and resource allocation are just a few examples. For example, dynamic programming agents have demonstrated the ability to play games like chess or Go at superhuman levels successfully. In autonomous navigation, tasks robots benefit from planning optimal paths using dynamic programming methods that optimize resource allocation—such as energy consumption—or computational resources used for critical application scenarios.

What is Dynamic Programming? In simple terms, it's a mathematical optimization method used to solve difficult problems by breaking them down into smaller subproblems, which can be more easily solved in manageable sections. This works on recognizing the principle of optimal substructure where solutions of sub problems solve higher order ones iteratively. Within Reinforcement Learning context specifically—it helps agents learn how to make decisions based on feedback received from their environments with one objective: maximizing cumulative rewards long term. In reinforcement learning, dynamic programming algorithms are indispensable tools. They operate by iterating backwards through a sequence of decisions to uncover the best option at each step along the way. Solving subproblems recursively and storing solutions in tables allows us later access for decision-making purposes, with the Bellman Equation being one popular algorithm leveraged for determining optimal policies.

The following algorithm is aimed at finding an optimal policy that yields maximum expected cumulative reward. These steps are iteratively repeated until the convergence of the policy. Although policy iteration works swiftly in simple environments, it may be computationally expensive in more complex ones. On the other hand, value iteration is a dynamic programming algorithm that solves reinforcement learning problems based on repeated value function updates determined by each state's expected cumulative reward under a specific policy until convergence.

Assuming we have a basic grid world problem with an agent moving from start to finish and each state representing a grid position, we employ value iteration to find the optimal policy that maximizes cumulative reward. First, we initialize every state's value function to zero and then update this function by considering all likely actions along with their corresponding rewards and resultant transitions until the values converge.

Let's consider an example where an agent is put into a small negative rewarding up left corner of Gridworld with only right and up movements possible—using the expected reward calculation methodology and updating the current cells value will produce numerical output. Once all states converge optimally for value functions, agents can decide upon their actions toward maximum cumulative gains.

Dynamic Programming encompasses diverse use cases comprising robotics applications like path planning/motion control/task execution/autonomous driving or game playing strategies optimization with decisions per turn/goal maximization techniques popularly used across industries worldwide.

When it comes particularly to path-planning and motion control within an obstacle-laden environment filled with multiple physical and practical constraints in place, dynamic programming can optimize complex and multifaceted decisions by calculating a cost-to-go function. The algorithm determines an optimal control policy that enables the robot to move along a desired trajectory while considering all dynamics and physical constraints.

Likewise, for task execution involving specific objectives, follow-through repetitive actions to achieve set goals requires good optimization techniques. These techniques are based on calibration of environment task matrices, robotic capabilities and constraints with an objective to compute an optimum policy for executing these tasks.

The potential outcomes are noteworthy. For example, a reason why dynamic programming excels is when optimizing robot movement within physical constraints while moving heavy objects in obstacle-filled environments, or in gaming AI applications where decisions in each turn based on heuristics algorithms could lead toward maximum impact in achieving overall goals such as winning chess. Playing chess requires one to strategize well while maintaining respectful demeanor toward your opponent as two players take turns to move their pieces on a board aiming at capturing their adversary's king.

In order to win, players need to have acute awareness of how their opponents might play alongside utilizing this information in making rational decisions which provides them with an edge over competition. One such approach utilized for making informed choices when competing in chess includes employing dynamic programming algorithms capable of scrutinizing available gameplay options during each turn. When it comes to autonomous driving, monitoring surroundings is essential for safe navigation on roads. The car's sensors can detect pedestrians, traffic signals, and other vehicles making it easier for a sophisticated system like a dynamic programming algorithm to keep up with real-time feedback. Likewise, this tool allows self-driving cars to have accurate predictions related to future events while adjusting speed or avoiding any potential accidents.

One of Dynamic Programming's main attributes is its ability to handle complex scenarios regardless whether they occur frequently or not. For instance, the algorithm considers various events such as sudden changes in climate or unexpected hurdles, hence deciding immediately what makes best sense.

Furthermore, policies Iteration and Value Iteration algorithms are often employed when solving reinforcement learning problems, due to their vast applicability encompassing gaming industry, robotics, and now automotive technology. During current times, Dynamic Programing continues to be a significant focus for research, promising in the future many more discoveries.

In conclusion, Dynamic Programming is a formidable tool that enables autonomous vehicles to make decisions efficiently. It breaks down complex tasks into smaller, more manageable subproblems, thus optimizing capacity as a self-teaching tool for engineering.

Monte Carlo Methods: Learning from Experience

Reinforcement learning involves utilizing various algorithms like Monte Carlo Methodologies to help get accurate estimates about what returns (rewards) different kinds of actions yield per simulation episode. Unlike other methods, this method does not rely on a particular environment model but rather trains from practical experience. How it works is that players sample random states and actions from various real-world settings and use them to better determine expected value for each state-action pair.

To show how Monte Carlo works, assume learning to play an arcade game—players keep playing different rounds, note down the rewards for each attempt. After playing gets repeated over many instances, the player can engage in informed guesswork about which action provides high stakes. Monte Carlo works by simulating countless episodes of similar tasks to establish accurate approximations on expected returns/overall reward on specific state-action pairs.

To calculate and estimate—returns for each individual that relies on averaging techniques across simulations specifically where an agent visits a particular state-action pair multiple times—after which the returns add up and get divided by the number of visits, then updated agent's policy based on these estimations (which have their limitations). An instance of such algorithms includes Monte Carlo Control, where results garnered simulate different goal-oriented scenarios which help derive optimal policies or make adjustments.

While simulation can be daunting—especially with complex problems—Monte Carlo has come through as efficient especially in significant games like backgammon or Go. The policy improves over time as this algorithm repeatedly takes actions with high hypothetical values until identifying an ideal policy. To illustrate, consider Monte Carlo Tree Search, employed in games such as chess and Go; it too efficiently assesses several possible future game states to calculate estimates for all moves' values. It selects actions with higher hypothetical values iteratively while continuing to simulate additional outcomes before a game concludes. In essence, Monte Carlo methods are renowned reinforcement learning algorithms that depend on numerous episode simulations in estimating expected returns for every action-state pair. Though resource-intensive, they remain effective at precisely identifying optimal policies for different tasks and are prominently used in robotics and gaming applications.

Temporal-Difference Learning: Combining Monte Carlo and Dynamic Programming

Reinforcement Learning provides tremendous opportunities for autonomous agents to acquire decision-making capabilities grounded in experiential knowledge. However, optimal solutions for such problems can prove expensive computationally while demanding significant volumes of data; therefore reducing both speed and scalability concurrently. To mitigate these limitations, we have a valuable solution: Temporal-Difference (TD) Learning Algorithm—where we merge both Dynamic Programming and Monte Carlo Methods strengths producing state-value predictions updated instantly rather than waiting until all episodes clean up! This property not only increases computational efficiency but significantly lowers time overhead as well.

The Q-learning algorithm is an effective method for convergence with optimal policies provided that all possible state-action pairs are explored by agents; however, balancing exploration-exploitation can be a challenge in practice. On-policy TD Learning algorithm SARSA updates value estimates based on reward-based selections achieved through a defined policy from both current and next states as well as actions selected accordingly.

SARSA proves successful due to its design for optimization even where exploration is required simultaneously within an environment. Exploration-enabling reinforcement learning applications such as robotics and autonomous vehicle control benefit from this approach resulting respectively—to teach robots puzzle solving or maneuvering obstacle courses—and similarly used for training self-driving cars how best to navigate traffic on highways or city streets.

TD learning exhibits strengths providing computational efficiency that facilitates direct real-life experience resulting in success when applied across diverse practical domains like game playing reinforcement efforts. In summary, TD learning solves reinforcement problems efficiently by integrating DP and Monte Carlo benefits.

Approximate Dynamic Programming: Scaling Reinforcement Learning to Large Problems

Reinforcement learning shows potential in solving various problems ranging from game playing to robot controlling. However, traditional reinforcement learning methods may have limitations when dealing with real world problems that are extensive and complex. To tackle this issue, approximate dynamic programming (ADP) algorithms use function approximation to enhance the scalability of reinforcement learning for broader problems.

ADP techniques incorporate ideas from both dynamic programming and function approximation by using function approximation to estimate the optimal value function instead of calculating it exactly as in dynamic programming. This way, we can mitigate issues arising due to a potentially huge number of states in the problem domain. In precise terms, function approximation involves utilizing a parameterized function to estimate the value function while minimizing errors generated by prediction values as compared with actual values stemming from problem domains. The technique empowers us to gain insights into underlying problem structures ensuring desirable predictions even for states never observed during training stages.

In particular, neural networks are an effective method for approximating functions utilized by ADP techniques since they can learn complex and nonlinear functions. In terms of categorization, ADP algorithms mainly come under two approaches: online and offline strategies that differ based on how they update the value functions. Online methodologies frequently update their estimates after interacting with the environment in small state domains quickly explored during implementation stages. On the other hand, offline approaches update their estimates using data batches collected outside of direct explorations and efficiently handle larger state spaces incurring computational benefits. One popular offline method for ADP is fitted Q iteration (FQI). The FQI algorithm's efficient iteration process builds datasets from collected environmental samples to update the estimated 'value' data effectively. Employing neural networks helps estimate this desirable quantity with parameters updated via regression algorithms such as 'least squares.' The alternate approach called 'approximate linear programming' (ALP) represented 'value' via functional approximation while solving a set of equations through linear programming methods to arrive at an approximate optimal solution yielding results almost similar to more common techniques.

Reinforcement learning applications have met success in various domains, be it robotics, finance, or energy management sectors. Here, ADP has aided significantly by enabling optimization of profitable business practices like wind turbine deployment through determining optimal policies based on previously acquired knowledge about state-value via cleverly designed value networks and greedy actions. Trading strategies, too, have seen substantial improvement from ADP interventions with value functions and computing optimal trade options using historical data to achieve similar gains in profitability.

The versatile algorithm family of approximate dynamic programming is ideal for scaling up the size and complexity of the problems tackled. Through function approximation techniques for estimating the state's value, this method encapsulates the underlying structure needed to make accurate predictions even when any given state has not been observed during training time. This powerful technique has a multitude of potential applications in various industries.

Policy Gradient Methods: Directly Optimizing the Policy Function

Policy Gradient Methods belong to a significant class of Reinforcement Learning algorithms specifically targeted at enhancing one critical aspect: direct optimization of policies mapping states to appropriate actions—commonly referred to as 'policy functions.' While alternate approaches require estimation of value functions with derived policies in tow, Policy Gradient Methods stand apart by exclusively concentrating on improving input-output mappings themselves using gradients derived from stochastic functionality evaluations using probabilistic methodologies filled with Monte Carlo sampling with no supervision needed whatsoever! Its versatility is harnessed through its applications in game-playing, robotics, and natural language processing. When it comes to solving intricate problems, deep reinforcement learning along with policy gradient methods showcase their sheer power via groundbreaking discoveries made using them.

Robots today possess the capability through reinforcement learning by receiving feedback or rewards from sensors at specific time intervals during operation thereby enabling them to master complicated motor skills in robotics such as grasping objects. Similarly, in natural language processing, machines today have the capacity to generate human-like language when modeled through policy gradient techniques. These techniques help in building a reward model during learning so that the machine can understand and then provide results accordingly.

However, it's crucial to note that these methods come with their own set of challenges. The high variance of policy gradient estimates proves to be one of their biggest obstacles resulting in slow convergence and unstable training, which can be rectified by using baseline subtraction and trust region optimization techniques. Striking a delicate balance between exploration and exploitation is yet another challenge faced while learning policies—too little or too much exploration could mean prematurely converging onto a local minimum instead of achieving an optimal solution.

Despite these challenges, new variants such as Vanilla Policy Gradient, Trust Region Policy Optimization, etc. made it feasible for algorithms today to work with complex state-action spaces encountered while dealing with large sets such as large dimensions. Vanilla Policy Gradient stands out for its incredible ability to train agents when it comes to optimal policy implementation whilst being guided by feedback datasets for each action taken through the reinforcement learning mechanism.

In order to train an agent to play a simple game involving navigating through a maze and collecting coins along the way, one could employ VPG methodology. Beginning with initializing a neural network representing the agent's policy that outputs probabilities for certain actions based on inputs from current game state, Monte Carlo estimation is used to approximate expected returns based on many simulated episodes (with corresponding rewards). By calculating gradients with respect to these expectations and updating iteratively toward optimal solutions, VPG provides both efficient usage for small problems as well as ease of implementation. However, caution should be taken when estimating gradients due to possible high variances leading to slow convergence or instability.

Introduced by OpenAI researchers in 2017, Proximal Policy Optimization (PPO) has gained popularity among researchers for training agents in environments with continuous action spaces. The primary goal of PPO is to update policies efficiently while maintaining stability during the learning process toward maximizing cumulative rewards while regulating policy updates to avoid instability.

One crucial aspect of PPO is the utilization of a clipping function that restrains policy updates' size. To ensure controlled policy changes and avoid instability, the clipping function limits policy change to a certain amount, even if the new policy results in a higher expected reward. With entropy regularization, PPO encourages exploration of the action space by penalizing low entropy policies that always choose the same actions in the same states. Using a surrogate objective function as a lower bound on the actual objective function allows PPO to optimize more efficiently and avoid large policy updates. Successful application of PPO includes training agents for robotic manipulation, game playing, and autonomous driving. However, high variance in estimated gradients can hinder learning an optimal policy when data is limited and the policy is complex.

Effective training of robots to navigate mazes requires addressing challenges that arise due to variable rewards for similar actions performed from identical locations in the maze. To address this issue, techniques like using baselines or trust region methods exist that can assist in stabilizing training processes and improve overall performance. While these are helpful techniques, choosing an appropriate learning rate may prove challenging as an incorrect one could lead agents toward suboptimal policies due to instability or slow convergence rates caused by excessively high/low values. Possible solutions include selecting from various learning rate schedules that adapt dynamically over time depending on changes in agent environment. Conversely, when setting a low learning rate, there is a potential for prolonging the robot's capacity building to navigate through the maze. Essentially increasing wait time before achieving desired outcomes. Subsequently, this can make it susceptible to arriving at an impasse or deviating from finding optimal navigation pathways.

Policy gradient methods have shown great success across a variety of tasks such as robotics control, game playing, and natural language processing. An excellent illustration of this is the story of AlphaGo, the computer program that defeated the world champion in the game of Go. AlphaGo Zero utilized a policy gradient method variant to achieve this feat.

The tale of AlphaGo is marvelous in showcasing the prowess of artificial intelligence and its ability to excel over humans in complex tasks. The journey commenced with Google's DeepMind developing a computer program called AlphaGo in 2016 for challenging Lee Sedol, a renowned South Korean Go player to compete in a best-of-five match. Go is an arduous board game with more possible moves than there are atoms in the observable universe—making it an exceptionally challenging game for computers to master. Yet AlphaGo was not just any ordinary computer program—it was an AI system trained via deep reinforcement learning techniques.

The event took place in Seoul, South Korea and millions worldwide watched it unfold. Sedol was considered one of the world's best Go players, having won 18 international titles and holding the highest rank possible within the game. The first match on March 9, 2016 provided quite a thrilling contest where both players encountered several surprises from each other's plays—with Sedol ultimately claiming victory. However, during their second round on March 10th—AlphaGo astounded viewers as it comprehensively defeated its human challenger by employing unusual yet creative moves never before seen within Go's history—winning through play after 176 moves instead of resignation. AlphaGo continued its dominance when they played their third match on March 12th—sealing another win for themselves against their expert opponent Sedol.

During their last encounter, AlphaGo showed off an exemplary performance that was artistically superb, incorporating innovative strategies with several unpredicted moves. Sedol struggled greatly but eventually had no choice and had to resign after only 176 moves. This victory by AlphaGo over Sedol highlighted the significance of artificial intelligence and its capacity for surpassing human capabilities in various tasks requiring expertise and in-depth understanding like Go playing technique.

Furthermore, many areas have experienced a drastic transformation thanks to technologies like policy gradient methods that have expanded into diverse industries ranging from medicine diagnosis to stock market prediction. These techniques have delivered positive results; however, they're not without limitations such as getting trapped at local optima or difficulty exploring state space resulting in restricted exploration capabilities not optimized for finding feasible policies.

Deep Reinforcement Learning: Combining Reinforcement Learning with Deep Neural Networks

Recently, there has been an exciting surge in advancement within the field of Reinforcement Learning due largely in part to merging it with Deep Neural Networks technology principles, offering compelling advances far beyond anything previously observed. Narrowly focusing on how to combine the strengths of both Deep Learning as well as Reinforcement Learning, this section takes a deeper dive into fundamental principles as well as examinations of how differing approaches in combining these fields' impact difficulties commonly faced and future opportunities offered.

Beginning with the basics of deep learning, our focus then moves to using deep neural networks for approximating various central features in reinforcement learning such as value function and policy. Furthermore, challenges that accompany a more complicated application of combining both fields are delved into—specifically stability issues, exploration difficulties—with potential solutions explored.

Traditionally speaking, reinforcement learning relied on excessively large tables containing state-action pairs to represent policies and value functions. However, this method is impractical for continuous-state problems or those facing larger state variables, such as real-world applications-.Deep Neural Networks technology represents a revolution in this arena by providing capabilities for abstract conceptual organization creating highly complex representations of both state and action spaces; thereby guaranteeing enhanced efficiency in learning.

As an example; when training robots how to play table tennis via traditional methods, humans would have to define all possible meaningful features such as ball position or paddle placement etc. This approach invariably led to limitations given the fallibility inherent in human attention spans/ability regarding infinite details coupled with complexity inherent in sophisticated tasks. In contrast, if we employ deep neural networks, such agents can learn its representation/viewpoint about nuances such as spin on the ball or predicting opponent's movement—making more nuanced decisions based on context awareness.

One significant benefit of deep neural networks in reinforcement learning is their capacity to generalize to new scenarios. Whereas traditional methods would necessitate retraining for each new case, deep neural networks can employ their prior knowledge to novel circumstances. For example, a table tennis robot may transfer its skills to playing against a different opponent or in another environment. Moreover, deep neural networks are equipped to deal with large and continuous state and action spaces.

Traditional approaches would find this unfeasible. Suppose a self-driving car must navigate complicated traffic scenarios with many variables, such as the position and velocity of other cars, pedestrians, and traffic lights. In that case, deep neural networks learn how to process this massive amount of information in real time and make decisions accordingly. In 2013, Google's DeepMind introduced the Deep Q Network (DQN) algorithm as one of the earliest examples of successful DRL algorithms available today. The algorithm incorporates deep neural networks' power with effective reinforcement principles such as Q learning algorithms.

By approximating the optimal action value function $Q(s, a)$, which forecasts expected reward levels by taking action 'a' in state 's' DQN is capable of learning a function that selects the highest valued actions available for an agent's corresponding state within any context it has learned. Even if it lacks experience taking actions within that precise context, DQN automatically learns useful features from raw input without requiring any engineering features.

Moreover, one critical benefit that DQN offers is its capacity to handle high dimensional input merely by using raw pixels from video games like Space Invaders without requiring feature engineering project tactics. Lastly, through implementing a replay buffer method during training when sampling mini batches of random experiences stored previously by an agent across time steps, DQN exploits experience to learn fruitfully for improving upon itself over time scales indefinitely! Preventing agents from overfitting to any particular experience requires exposing them to diverse sets.

Directly mapping current states of environmental actions via policy functions provided powerful solutions ideal for handling continuous action spaces when compared with Q-learning approaches' discretization requirements. This explains why popular policies such as deep deterministic policy gradient (DDPG) algorithms find success easily since they remain faster while avoiding significant computational burdens common in traditional policy gradient techniques like Q-learning. In games like Atari, deep Q-networks (DQN) were capable of surpassing human-level performance by detecting playing strategies based on raw pixels while learning from a diverse set of experiences in the robotic field. Here DQN taught robotic arms how to perform tasks like picking objects and placing them at target locations. However, it's important to note that DDPG becomes an efficient policy gradient architecture ideal for discovering continuous control policies for high-action dimensional spaces. It builds upon DQNs that were initially suitable for discrete actions spaces. The agent's main task involves observing the present state of its environment—specifically, the arm's position and the target. After careful consideration, it then selects an appropriate course of action to take—usually by determining the arm movement angle. By aiming to learn policies that maximize reward, the agent works toward successfully reaching its intended goal.

DDPG functionally relies on two neural networks, namely: an actor network and a critic network. The former analyses current environmental readings provided as input and generates output actions for the agent to perform. The latter receives both readings about present environment status and any actions performed by the agent as input, generating an anticipated reward value estimate accordingly.

DDPG training improves as it uses these neural networks to make more informed decisions about which steps to take next while estimating their expected rewards upon successful completion. As such, during training phases, agents use their actor network to pick actions that align with present environmental states while relying on critical networks for calculating reward estimates in consideration of such actions. Additionally, agents benefit from a form of policy gradient algorithm during network weight adjustment phases—further enhancing their capacity for maxing expected rewards.

DDPG boasts advantages over other methods when it comes to handling continuous action spaces like those used in robotics or autonomous driving applications where fine control is necessary for picking up objects or effectively navigating busy roads respectively. Also notable is DDPG's stability and sample efficiency stemming from its ability to rely on two neural networks dedicated environments, assessing atmospheric data while also estimating policy and value functions more efficiently than other methods.

The Trust Region Policy Optimization (TRPO) algorithm is widely known for its effectiveness in policy gradient methods. It is frequently utilized in the field of Deep Reinforcement Learning to enhance the stability and efficiency of these methods. The primary challenge that such methods often face is the struggle to strike a balance between exploration and exploitation while avoiding getting stuck in local optima. To gain an understanding of how TRPO operates, let's consider an example where you aim to impart knowledge on a robot about navigating a maze. The robot has access to a camera that enables it to observe its surroundings, while also having the ability to take actions like moving forward, turning left or right. The objective is based on finding a suitable policy that optimizes the likelihood of the robot reaching the end of the maze. In conventional policy gradient methods, neural networks are employed as policies that receive information from their environments, applying outputs onto probability distributions with potential actions.

This network utilizes stochastic gradient ascent which implements variants of gradient descent, working toward maximizing expected rewards over possible trajectories within reach for said robot. However, one finds several limitations with this model:

issues with sensitivity to learning rates leading to suboptimal policies deviating or exploding during training; high computational costs for Monte Carlo method gradients; and challenges concerning updates size stability and management strategies.

TRPO approaches this by implementing trust regions which limit the size of policy update ensuring little deviation from current policies. By valuing nearness versus guaranteeing reward maximization in comparison with previous policies, TRPO's trust region are defined enough toward stable improvements without risking overall performance degradation through overlarge changes. This is solved via constrained optimization problems optimizing expected reward outputs under applicable distance metric constraint updates accordingly mirrored therein.

The efficient exploration of new paths without disturbing robots' ongoing behavior is crucial for learning processes that can escape local optima. This is done while optimizing policy rewards using deep reinforcement learning or DRL algorithms such as TRPO which use surrogate objective functions optimized by conjugate gradient descent calculations without requiring expensive second-order derivative calculations. Examples of TRUMP's success range from OpenAI's humanoid robot training for performing complex movements like running and walking to training agents playing Atari games with outstanding performances exceeding human capabilities. Other DRL algorithms like PPO are known for their effectiveness in manipulation tasks while A3C boasts of state-of-the-art achievements in gaming tasks.

Challenges of Deep Reinforcement Learning

The effectiveness of deep reinforcement learning depends on how efficiently datasets are used during neural network training. Unfortunately, this often requires substantial investments both in time and resources making broad applications such as developing AI capable of playing chess at an expert level challenging.

Different solutions have been proposed toward addressing this hurdle including Transfer Learning—whereby trainee algorithms are taught simpler games like Tic tac toe before advancing onto more complex ones. Another approach involves Generative models which create synthetic datasets that resemble actual scenarios—significantly reducing the amount of real world data required to train algorithms.

But overfitting remains a potential issue in deep reinforcement learning. It limits model adaptability to new environments beyond known boundaries affecting the algorithms' wider applications. Therefore, besides avoiding over reliance on vast datasets during training, attention should also be directed toward ensuring that outputs are generalizable outside familiar scenarios. In the field of robotics, training a robot to navigate a maze using reinforcement learning is a common practice. In such cases, the robot is rewarded for successfully reaching the end of the maze and penalized for crashing into walls.

However, if the robot's neural network model becomes too complex during the training phase, it may over memorize its path through the maze rather than formulating policies on how to navigate any maze in general. This will have implications on its performance when tested with new mazes. Several techniques can be used to avoid this issue and some of them include regularization techniques like weight decay or dropout, which encourage better policies instead of overly complex ones by incorporating penalties in loss functions. Early stopping can also be used to halt training when performance drops while experience replay and data augmentation can help increase available training data.

Reinforcement learning agents access their environment for information and continuous interaction occurs over time. Underlying patterns that explain changes in an environment are referred to as non-stationary environments where these changes are not random or stochastic. Researchers face a significant challenge dealing with non-stationary environments in deep reinforcement learning because they require fast adaptation from agents to maintain their high performance levels. A common example encompasses factory floors that robots navigate making use of dynamic processes as machinery is added or removed due to system expansion requirements enhancing production.

To deal with non-stationary environments, various research techniques have been developed. One such method is called "online learning" where the agent updates previous experiences with new information policy alterations that occur dynamically keeping pace with changing patterns. This ensures high performance levels throughout time lapses. In industrial settings such as factories, robots must navigate through various obstacles while performing various tasks efficiently.

One way robots excel at this challenge is by adjusting their navigation strategies instantly when confronted with new obstacles instead of waiting till the end of a task completion cycle before updating them. To ensure better memory retention of previous experiences, another method called memory storage allows robots to store past experiences that provide guidance on future decision making processes. Reward signals play a critical role in reinforcing desirable behavior in reinforcement learning systems; hence developing accurate reward functions remains necessary but can be difficult at times. Reward shaping provides an innovative technique that modifies existing rewards or adds new ones for enhancing agent behavior performance. This approach applies new pathways toward successful completion of designated programming goals.

In certain task situations such as navigating mazes or playing games like Super Mario Bros., designing specific incentive policies for individual actions provides excellent metadata feedback for agents to evaluate their success rates accurately. Reward shaping can harness these unique situations to add extra incentives or modify existing ones that agents need to accomplish their assigned tasks efficiently.

Robotics is another field where reward shaping has been useful especially in autonomous robots. Robots tasked with specific sorting tasks can get higher reward incentives for correctly sorting items; however, putting a wrong item in the right box may attract low rewards. Though our initial rewards system has helped guide our robots toward improved object sorting protocols thus far, there are limitations as we strive for more efficient practices. To this end, including supplementary cues, which incentivize speedy sort methods and conserving power reserves via model "reward shaping" can lead us down more effective paths.

Nevertheless, it's essential to approach reward shaping with caution, as there are inherent challenges in designing suitable supplementary reward systems. With so much at stake, poorly designed functions can drastically impede our progress, or in extreme cases, even contribute negatively to our robot training efforts.

One challenging aspect of determining the ideal system lies in potential biases that new rewards could risk injecting into the learning process, thus affecting how well we can apply learned policies to other realms.

Henceforth, it is essential to conduct detailed assessments while shaping incentives that work without attaching unintended risks to our robotic solutions.

Model-Based Reinforcement Learning: Learning the Dynamics of the Environment

When it comes to optimizing long-term rewards through reinforcement learning algorithms, utilizing model-based strategies comes with significant advantages over traditional methods predicated on trial-and-error exploration alone. Constructing accurate behavioral models is based on agents' observations and specific features relevant for reward optimization. Mapping state transitions between actions performed and rewards received, which is recorded within those environments offers efficient ways for improved learning while minimizing required interactions with them at evaluation time or application phase.

Improvements in sample efficiency result in better agent proficiency, which leads to improved long-term performance outcomes. Notably, using a Model-Based Approach has several advantages over Model-Free Approaches especially when planning is required as demonstrated through robot navigation in mazes. In this case, building a model of the maze based on information such as walls layout allows for simulation of various schemes resulting in better planned paths—hence efficient navigation with fewer constraints and reduced time consumption.

Transfer Learning between tasks comes with immense benefits especially where acquiring sufficient training data would have been otherwise impossible. Producing an environment model provides shared knowledge across similar scenarios. Here, experience from previous-related tasks can be utilized for gaining insight into how best to accomplish new goals leveraging on pre-existing knowledge for effective use of time while having high chances for mission success. When faced with new tasks or environments, building off previously gained knowledge and expertise can prove valuable for agents in their ongoing learning processes.

For example, consider how robot operators use currently possessed skill—like successfully navigating previously known factory floors filled with obstacles—during training exercises this tool is called "experience." Reinforcement learning heavily emphasizes discovering ways that an agent might navigate between exploring these environments first hand versus immediately attempting to apply previous learned experience right out. This tension defines one's exploration/exploitation trade-off experience, especially when resources will be scarce compared to ideal scenarios. Thankfully, using MBRL consistently enables us to address some of these challenges by allowing agents access tools such as maps—aka models of particular environments (mazes etc.). Imagine it is lost in a premiseless labyrinth without any guidelines. In an instance like this, one might zigzag through random paths in search of solutions using a "hit or miss" tactic—employing diverse plans which do not always have the desired effects.

Increasingly, agents can apply MBRL strategies to better pre-determine optimal routes with regards to exploration and end-goal attainment (such as exit). One primary issue that arises in using MBRL is developing accurate models of the environment before it can all be learned well. Striking a balance between model complexity, data availability, and desired level of generalization is necessary for optimal decision making in Machine Based Learning Recommender (MBLR). Overly simple or inaccurate models lead to suboptimal decisions while overly complicated ones overfitting on available data result in poor generalization to new scenarios.

Deep neural networks present an accurate approach for learning complex non-linear relationships between inputs and outputs at MBLR environments with high-dimensional state-action spaces like Go's game states. Here, AlphaGo attained champion status using a value function approximated by a deep neural network.

Ensemble methods provide powerful techniques for combining multiple machine learning algorithm strengths resulting in improved performance at MBLR contexts like cold-start problems, sparsity, or scalability issues facing traditional recommender systems. Model averaging ensemble type consisting of training several models on subsets of data like varying time periods or genres and averaging their predictions offer a promising avenue toward generating more accurate, diverse recommendations at MBLR. Another approach to ensembling in MBLR is through model stacking ensembles.

This entails training several models and taking advantage of their forecasts as variables for an advanced-level model. To expound further, if multiple movie rating prediction models exist based on different features like director, cast or genre—these predictions can serve as inputs to a more sophisticated model that consolidates these forecasts for enhancing final recommendations.

Ensemble learning bears significance in MBLR, since it permits us to benefit from multiple models' strengths while neutralizing their drawbacks individually. By amalgamating numerous models via this method, we can augment accuracy and variety in our recommendations alongside decreasing overfitting or bias risks effectively. However, essentiality lies in carefully choosing suitable constituent members for this approach comprehensively evaluating them before inclusion so they remain relevant beyond existing data sets.

There are several **benefits to employing MBRL** including better sample efficacy and transfer learning. However, it is crucial to acknowledge existing issues behind this methodology which includes, among others, potential problems with accounting for both 'model bias' or 'model mismatch'. Model bias arises when an agent uses an inaccurate representation of environmental conditions within its modeling processes—in other words, incomplete/inaccurate data sets may be examined leading to poor decision making on behalf of agents navigating through instances like mazes etc… Some researchers within this field propose resolving these issues by employing more complex models that can encapsulate environmental conditions better, or with the usage of more varied data sets. Model mismatch is when an agent uses a model that doesn't replicate environmental dynamics leading to anomalous/thwarted outcomes. Teams have been developing various techniques for mitigating model bias and model mismatch in MBRL including ensemble models, which combine multiple models to reduce the impact of individual model's biased analytics.

Model-based reinforcement learning presents another avenue where we can apply our models' uncertainties through estimation techniques for accurate decision-making by agents avoiding failures caused by overly confident choices. Model uncertainty comes into play when developing our MBRL predictive models despite providing accurate insights into its environment because factors like measurement error also induce uncertain elements leading to instabilities during training processes even when designed correctly. A classic example will be developing robots' ability to play games like chess through constructing a machine learning model reflecting perfect policies predicted taking into account all variables within the game. Yet, even with accuracy fitting perfectly well, limitations may arise resulting in the model becoming prone to predict incorrect moves that do not reflect the best possible outcomes. Predictions may have to include uncertainties such as possible limitations within the model's prediction accuracy, making less-optimal decisions for robots, leading to potential game losses.

For instance, in self-driving cars, predictions arising from models created using data on driving practices in various environments could become heavily uncertain due to environmental factors like weather conditions or unpredictable obstacles. Hence, they may cause drivers to make wrong decisions ultimately ending up causing accidents. The exploration-exploitation predicament constitutes another obstacle for MBRL. It entails striking a delicate balance between trying out new actions so as to gain insights into the environment and sticking to existing policies for maximizing rewards. This challenge looms larger in complex environments where exploring proves costly both in terms of time and resources.

Multi-Agent Reinforcement Learning: Learning to Cooperate and Compete

Reinforcement learning offers individual agents a potent tool for decision-making problems. Nevertheless, multiple-agent interactions are required for many real-world scenarios when assessing ways in which they can interact effectively with each other. This demands Multi-Agent Reinforcement Learning (MARL), which explores situations where many agencies optimize their decisions via self-learning mechanisms while targeting shared or independent objectives.

Two primary approaches exist under MARL - cooperative and competitive methods.

Cooperative MARL entails various agencies collaborating in a shared environment aiming at collective achievements despite different responsibilities under specific tasks. A perfect example is robots assembling a car where different robot agents handle specific tasks but must work collaboratively to ensure that the car assembles accurately within optimal timescales. Under Cooperative MARL, every agent has access to other agents' action and observation data crucial for making informed decisions toward achieving mutual goals.

Cooperative MARL presenting several advantages over individual reinforcement learning such as enhanced productivity when addressing complicated tasks coupled with peer-sharing opportunities of experiences. It poses diverse challenges under Unique balancing aspects including competition versus cooperation approaches that requires addressing. In addition to this is preventing communication breakdowns from redefining objectives midway during engagement and mitigating possibilities of free-riding whereby some agents may not contribute toward successfully achieving shared objectives.

The field of Multi Agent Reinforcement Learning (MARL) involves various agents competing against one another in shared environments whereby they learn over time how best they can achieve their individual goals without obstructing their competitors' quest for success too much. Competitive MARL thus necessitates striking a delicate balance between maximizing one's reward while preventing others from doing so too much because whoever wins gets the spoils. For instance, during a game of poker, every player desires winning over everyone else by either having an unbeatable hand or manipulating opponents into conceding defeat (folding). However, this also requires them to consider what others may hold and realize that nothing is assured until the final cards get revealed. Alternatively, when autonomous vehicles operate on a crowded road with others competing for similar destinations, their goal is to get there quickly and efficiently while also avoiding any accidents with other cars.

The complexity of competitive MARL arises from the non-stationarity of the environment in which agents compete as each agent's actions change how every other group member perceives and acts in the environment. Consequently, ensuring effective strategies requires techniques like deep reinforcement learning, actor critic methods, self-play mechanisms or opponent modeling that adapt to such dynamic scenarios. Non-stationarity is a significant challenge experienced in Multi-Agent Reinforcement Learning (MARL) as various changes occur due to several factors such as changes observed among other agent behaviors or modifications done on game rules and environment settings. A suitable instance that supports this complexity is exemplified through considering a soccer match played by a group of robot players trained explicitly on scoring goals collectively and impeding opponents' scoring success rates.

The opposing team's conduct might turn unpredictable over time with sudden displays of aggressiveness, defensiveness, or implementing new tricks altogether to win the match. Adapting to these transitions, and maintaining team effectiveness becomes arduous.

Besides the soccer scenario above, other MARL cases suffer non-stationarity. For instance, market conditions in a multi-agent trading setting shift rapidly and affect agent behavior that may result in incorrect predictions and subsequently wrong decisions. Research has seen various approaches developed to handle such situations through ensemble methods where different models train on diverse data subsets from the environment or deploying meta-learning skills allowing agents to learn quickly and adapt optimally.

Moreover, successful credit assignment presents another major challenge faced when dealing with Multi-Agent Reinforcement Learning (MARL). The problem arises when rewards received among several players interacting with different environmental settings become challenging to distribute fairly among respective players present during interaction. Imagine a soccer match with multiple players who act as agents in MARL. The team's goal is to score as many goals as possible while preventing the opponent from scoring any shots at goal. Since each player has different roles during playtime, determining which individual was crucial in achieving victory can become difficult.

In particular, assigning credit for rewards also becomes more complex when one's objectives differ from another team member's aim. Imagine if some members strive for speed while others prioritize energy efficiency, both working toward a common objective; allocating merit is hence paramount in ensuring that incentives encourage optimization of relevant parameters.

Scalability presents notable challenges in Multi-Agent Reinforcement Learning (MARL). Over time, more participants join an environment in which interactive relationships might be developed among them. This could cause learning a decision-making pattern aggregating both beliefs and expectations while optimizing each participant's cumulative benefit that becomes burdensome work, unsolvable by any human means alone. Imagine constructing a group of robots working collectively toward achieving shared goals like manufacturing objects or conducting disaster response efforts. In order for their plans to succeed efficiently, they must act cohesively. Nonetheless, due to granularity margins which relate changes among interacting recipients, the problems could become intricate when the numbers increase dramatically.

As such, it is demanding to develop accurate models regarding these processes whilst estimating individual value computations. As much as an environment becomes more excessive concerning participants, the pattern of interactions could become more high-dimensional and changes are difficult to predict accurately. One huge issue concerns how to redistribute efforts while training several participants effectively. This inclusive factor makes collaboration challenging among agents, with `artificial intelligence (AI)` spread unevenly for more concise impact. Deep learning has been used alongside Multi-Agent reinforcement learning extensively across various domains including robotics, gaming and autonomy. What makes this collaboration so fruitful is that deep learning provides an excellent tool for representation learning while function approximation. MARL enables multiple agents to learn from both cooperation and competition thereby adaptively learning from their surrounding environment.

One example of such application is in autonomous driving, which requires multiple agents such as cars, pedestrians and traffic lights to dynamically behave in a complex environment. Researchers use convolutional neural networks (CNNs) and Recurrent Neural Networks (RNNs), which efficiently process sensory inputs (camera, LIDAR or GPS data) by selecting features relevant to complex decision-making processes. Following this, agents can apply MARL algorithms like Q-learning and Policy Gradient Methods to accumulate insights from the interactions among themselves thereby learning how to maximize reward—making them more intelligent and efficient over time.

In robotics domains, MARL coupled with deep learning can model diverse robots coordinated in achieving a common goal like assembling a product or exploring an unknown environment. DNN and RL help these robots to learn communication skills including perceptions and actions with dynamic coordination based on accumulated knowledge about their environment. The algorithm used here includes Actor-Critic methods and Model-Based Reinforcement Learning aimed at providing optimal ways for the coordinated actions of team members.

Lastly is how deep learning in tandem with MARL has found its way in gaming, especially competitive multiplayer games like Dota 2 or StarCraft II where different players control their unique agents and compete against each other within a complex environment. Here, CNNs and RNN can be utilized to obtain valuable game state information, which will predict opponents' anticipated actions. On the other hand, MARL algorithms such as Multi-Agent Deep Reinforcement Learning and Hierarchical Reinforcement can optimize agent behavior by safely coordinating their individual move toward ensuring victory for their team. Apart from several benefits promised by Multi-Agent Reinforcement Learning

(MARL), one functional area shows great promise—its application to autonomous vehicles.

This comes amidst demands from industry leaders moving toward self-driving transportation infrastructures; requiring smart systems capable of interacting among themselves effectively to optimize traffic flow avoiding congestion or collisions where possible. With emerging applications suitable for complex dynamic system environments such as autonomous driving operations is where MARL shines bright—making it a very promising approach toward effective decision-making across different sectors including autonomous transport systems.

It's no secret that harmonious interactions between multiple decisions made by independent actors within the environment find wide-scale applications beyond just autonomous transport systems. In the context of self-driving vehicles, MARL can effectively resolve various decision-making challenges such as traffic management, route planning and coordination necessary between individual vehicles to enhance safety protocols in congested environments. Picture multiple self-driving vehicles traversing busy highways communicating with each other for safe navigation while cautiously maintaining effective speeds, an excellent example of MARL's functional solutions.

A practical example such as Cooperative Adaptive Cruise Control (CACC) system showcasing the advantages of MARL within autonomous systems models automobiles that communicate among themselves through wireless networks designed to maximize driving safety and efficiency. Each car maintains a suitable distance from its predecessor using in-built communication networks that constantly update all vehicles' information to allow coordinated movements for efficient traffic flow management while assisting with enhanced safety standards.

Another instance is Intersection Management System that coordinates autonomous automobiles to navigate through intersections safely by conducting internal negotiations between cars based on diverse parameters entered into its model—thereby offering a functional solution for smoother intersection crossings.

Real-World Applications of Reinforcement Learning

In recent years, reinforcement learning has made significant progress and is increasingly being applied to real world problems. This section discusses some of the most notable applications of reinforcement learning. One of the most exciting areas where we see this technology emerging is in robotics. With developments in technology, robots are becoming more widely used across a range of settings including manufacturing plants and homes. Thanks to deep neural networks and other modern machine learning techniques, researchers have developed robots that can perform various complex tasks in dynamic and unpredictable environments. One of the key advantages of reinforcement learning for robotics is that it allows these machines to learn and adapt to new situations without explicit programming. This significantly improves their ability to deal with unexpected obstacles or changes in the environment—such as those encountered during logistics or manufacturing tasks.

An excellent example of using reinforcement learning in robotics is through developing robots that can autonomously navigate complex environments. For instance, researchers have created robots that can navigate through cluttered rooms, avoid hurdles and even climb stairs—all by utilizing reinforcement learning algorithms. The machines are capable of adapting their behavior to different scenarios based on prior experiences without needing human intervention.

Reinforcement learning has also proved useful for developing robots that can perform more specialized tasks such as object manipulation or grasping with high levels of accuracy and reliability. Researchers achieved this feat by training these robots using simulated environments before fine tuning their behavior through reinforcement learning.

Autonomous vehicles represent another crucial area where we find reinforcement technology at play—researchers utilizing RL algorithms to teach cars how to navigate complex traffic environments while making decisions in real time, paving the way for a future where self-driving cars become commonplace on our roads.

Lastly but no less important, the application area is healthcare—the use of RL healthcare includes clinical decision making, disease diagnosis, treatment planning, and drug discovery optimization protocols personalized, based on individual patient characteristics (medical history). Reinforcement learning has become increasingly pivotal in healthcare's advancement field through deep reinforcement learning methods that analyze medical imaging accurately. Precision diagnosis and illness detection provide extensive promise for detecting cancer at early stages. Utilizing RL algorithms improves already high efficacy metrics, positively aiding certain illnesses amongst patients better individually given unique medical histories and genetic makeups.

Clinical decision making utilizing reinforcement technology offers valuable modern insights into how personalized treatments for patients are feasible. By utilizing patient data to train the reinforcement algorithm, treatment decisions that account for specific medical history and current condition are made effectively and efficiently, potentially decreasing healthcare costs while improving various patient outcomes.

RL in drug discovery offers researchers a faster and more efficient design process whereby tailored algorithms facilitate new drug development for specific diseases decreasing associated costs.

Additionally, RL also revolutionizes transportation. As exemplified by self-driving cars' complex decision-making abilities, through deep RL capabilities, vehicles can learn to navigate challenging scenarios such as traffic conditions or light signals accurately. The results of which are a safer driving environment with optimized traffic flow that decreases hassles from drivers and new safety advantages.

Further, transportation optimization can be achieved via intelligent traffic control systems that use RL to optimize diverse routes globally like speed limits or vehicle rerouting efficiently. This reduces travel times worldwide enhancing logistics network advancements within different fields of industry overnight. Efficient handling of operations by DHL is possible with the use of reinforcement learning—an optimization approach.

Through perfecting their delivery routes with this strategy, they have been able to minimize fuel utilization as well as duration spent on deliveries. This innovation can also be applied in scheduling transport vehicles including planes, buses or trains, allowing maximum output in terms of time-effectiveness.

MACHINE LEARNING EVALUATION

Introduction to Evaluating Machine Learning Performance

In the exciting world of machine learning, we create smart algorithms that learn from data and make predictions or decisions. But how do we know if our algorithms are doing a good job? That's where "Performance Evaluation" comes in!

Imagine you have a robot that can classify whether an animal is a dog or a cat based on pictures. You want to check how well the robot can do this, right? That's exactly what performance evaluation is all about—finding out how good our machine learning models are at their jobs.

In simple terms, performance evaluation helps us measure how accurate and reliable our machine learning models are. It's like grading the robot's ability to tell dogs and cats apart. This helps us choose the best algorithm for a specific task and understand how confident we can be in its predictions.

We don't want our robot to make too many mistakes, right? Just like that, we don't want our machine learning models to make too many errors when predicting. So, we use different metrics to understand their performance. These metrics are like report cards for the models, telling us how well they did.

In this chapter, we'll explore various metrics and ways to evaluate machine learning models. We'll learn how to measure accuracy, which tells us the percentage of correct predictions. We'll also dive into other metrics that help us understand things like how often the model correctly identifies positive cases (like sick patients) or how well it ranks items.

The Importance of Performance Metrics in Machine Learning

When we build and train machine learning models, we want them to be as accurate and reliable as possible. Just like how we want our superhero to save the day flawlessly, we want our models to make the right decisions with confidence!

Now, you might be wondering, "Why do we need to bother with these performance metrics?" Well, let me explain it in simple terms.

Performance metrics are like guiding stars for our machine learning journey. They help us understand how well our models are performing and if they are making good predictions. Without these metrics, we'd be flying blind—not knowing if our models are doing a fantastic job or need some improvements.

Imagine we have a magic wand that can predict whether it will rain tomorrow. We would be excited to know how often it gets it right, right? That's exactly what performance metrics do for us— they tell us how often our models get it right and how often they might make mistakes.

Think of these metrics as magic spells that reveal the strengths and weaknesses of our models. They give us insights into what our models are good at and where they might need some extra training. Just like a coach helps a sports team improve their skills, performance metrics help us fine-tune our models to become better predictors.

Moreover, performance metrics are vital when we compare different models. It's like comparing different superheroes to see who is the mightiest. We want to know which model performs better, so we can choose the best one for our task.

By using performance metrics, we can make informed decisions about our machine learning models. We can spot potential issues early on, celebrate their successes, and, most importantly, ensure they are reliable when we use them to solve real-world problems.

In the realm of machine learning, the traditional statistical hypothesis testing methods that are commonly used in statistics do not directly apply. The unique nature of machine learning algorithms necessitates the development of specialized performance metrics for evaluating their effectiveness. Let's delve into why this distinction exists and why traditional statistical approaches fall short in the context of machine learning.

In statistical hypothesis testing, the primary objective revolves around understanding population parameters and drawing inferences based on a limited sample. Techniques like t-tests, ANOVA, and chi-square tests aim to determine the significance of relationships or differences between variables in a population. However, machine learning operates on a different paradigm altogether.

Machine learning algorithms are primarily concerned with learning patterns and making accurate predictions based on data, rather than making population inferences. Instead of estimating population parameters, the focus is on the model's ability to generalize and make reliable predictions on unseen data. Consequently, traditional statistical hypothesis testing methods do not provide direct measures of predictive performance, which is of utmost importance in machine learning.

One of the key factors contributing to the need for unique performance metrics in machine learning is the inherent complexity and flexibility of these models. Machine learning algorithms can be highly intricate, incorporating a multitude of features and accounting for complex interactions within the data. Such complexity poses challenges when attempting to utilize traditional statistical tests, which often rely on assumptions about data distribution or linear relationships between variables.

Furthermore, machine learning encompasses a diverse set of algorithms, including non-parametric models like decision trees, random forests, and neural networks. These models do not make strong assumptions about the underlying data distribution, making traditional statistical tests ill-suited for evaluating their performance.

In contrast to traditional statistical testing, machine learning models are evaluated using metrics that reflect their predictive capabilities and generalization performance. Accuracy, precision, recall, F1-score, and area under the curve are examples of such metrics that capture different aspects of a model's performance. These metrics are designed specifically to assess the model's ability to make accurate predictions, handle trade-offs between different evaluation dimensions, and leverage the information present in the data.

Additionally, machine learning necessitates a data-driven approach, where models learn patterns and relationships from the available data. This data-driven nature further emphasizes the need for performance metrics that effectively gauge the model's ability to capture and utilize the information inherent in the data.

In summary, the unique characteristics of machine learning, such as its focus on prediction, complexity and flexibility of models, non-parametric algorithms, multi-dimensional evaluation, trade-offs, and data-driven nature, require the development and utilization of specialized performance metrics. These metrics provide a comprehensive understanding of a model's predictive power and its ability to solve specific tasks effectively. By embracing these metrics, we can quantitatively evaluate and compare the performance of machine learning models, ultimately driving advancements and improvements in the field.

Classification Performance Metrics - Measuring Model Accuracy

In machine learning, one of the essential tasks is classification, where models are trained to predict different categories or classes. For example, we might want a model to distinguish between "spam" and "not spam" emails. But how can we tell if our model is doing a good job? Let's explore some simple ways to evaluate the accuracy of classification models.

Accuracy: Think of accuracy as a report card for our model. It tells us how often the model makes correct predictions. For instance, if our model predicts correctly 80 out of 100 times, the accuracy would be 80%. A high accuracy means our model is making more correct predictions than mistakes.

Precision: Precision focuses on how well our model identifies positive cases correctly. Imagine a doctor diagnosing a disease. Precision measures the proportion of correctly diagnosed patients among those the doctor classified as having the disease. We want high precision because it means fewer false positive predictions, reducing the chances of misdiagnosis.

Recall: Recall measures how well our model captures all the positive cases. Using the same doctor example, recall looks at the proportion of patients correctly identified as having the disease among all the patients who actually have it. We want high recall because it means we're not missing many positive cases, minimizing the risk of overlooking potential problems.

F1-Score: F1-Score combines both precision and recall into a single metric. It helps balance these two aspects. We want a high F1-Score to ensure that our model is both precise and able to capture positive cases effectively.

Confusion Matrix: The confusion matrix summarizes our model's performance. It shows the number of correct and incorrect predictions for each class. For a binary classification problem, it displays true positives, true negatives, false positives, and false negatives. This matrix helps us analyze where our model is making mistakes and where it excels.

ROC Curve and AUC: The Receiver Operating Characteristic (ROC) curve illustrates how our model performs at different classification thresholds. It plots the true positive rate against the false positive rate. The Area Under the Curve (AUC) summarizes the overall performance of the ROC curve. A higher AUC value indicates better model performance.

By using these classification performance metrics, we can gain valuable insights into how well our model is performing. We can assess its accuracy, precision, recall, and overall effectiveness in making correct predictions. These metrics guide us in understanding where our model might need improvements or adjustments.

Regression Performance Metrics - Assessing Model Accuracy in Easy Terms

When it comes to machine learning, one common task is regression, where models predict numerical values instead of categories. For example, we might want a model to predict the price of a house based on its features like size, location, and others. But how can we measure how well our regression model is doing? Let's explore some simple ways to evaluate its accuracy.

Mean Absolute Error (MAE): MAE tells us, on average, how far off our predictions are from the actual values. It calculates the average of the absolute differences between predicted and actual values. A lower MAE means our predictions are closer to the true values, indicating a better-performing model.

Mean Squared Error (MSE): Similar to MAE, MSE measures the average of the squared differences between predicted and actual values. Squaring the differences gives more weight to larger errors. Like MAE, a lower MSE indicates a better fit of the model to the data.

Root Mean Squared Error (RMSE): RMSE is derived from MSE by taking the square root. It brings the metric back to the original scale of the target variable. RMSE is commonly used because it is easy to interpret and shares similar characteristics with the mean.

R-Squared (R2): R-squared represents the proportion of the variance in the target variable that can be explained by the model. It ranges from 0 to 1, with 1 indicating a perfect fit. A higher R-squared value suggests that the model captures a larger portion of the variability in the data.

Adjusted R-Squared: Adjusted R-squared takes into account the number of predictors in the model. It penalizes the addition of irrelevant predictors, helping to prevent overfitting. Adjusted R-squared is useful when comparing models with different numbers of predictors.

Mean Absolute Percentage Error (MAPE): MAPE measures the average percentage difference between predicted and actual values. It is commonly used in business and forecasting scenarios. However, MAPE can be sensitive to extreme values and influenced by zero or small actual values.

By using these regression performance metrics, we can gain insights into how well our model performs in predicting numerical values. We can assess the average error, the spread of errors, the proportion of variance explained, and the percentage difference. These metrics help us understand the strengths and weaknesses of our regression model and make informed decisions for improvements.

Evaluation Metrics for Imbalanced Datasets - Assessing Model Performance with Imbalance

In machine learning, we often encounter situations where the distribution of classes in the dataset is imbalanced. This means that one class has significantly more examples than the others. For example, in fraud detection, the number of fraudulent transactions is usually much smaller than the number of legitimate transactions. When dealing with imbalanced datasets, standard evaluation metrics may not provide an accurate representation of model performance. Let's explore some simple ways to evaluate models in such scenarios.

Accuracy alone may be misleading: Accuracy, which measures the overall correctness of predictions, can be misleading when dealing with imbalanced datasets. If a model predicts the majority class most of the time, it can achieve high accuracy simply by ignoring the minority class. Thus, we need additional metrics to assess performance more effectively.

Confusion Matrix: The confusion matrix is a valuable tool for evaluating imbalanced datasets. It provides a breakdown of predictions into true positives, true negatives, false positives, and false negatives. These values allow us to calculate various performance metrics that provide a more comprehensive view.

Precision: Precision focuses on the accuracy of positive predictions. It measures the proportion of correctly predicted positive instances out of all positive predictions. In imbalanced datasets, precision helps us understand the reliability of the positive predictions made by the model.

Recall (Sensitivity or True Positive Rate): Recall, also known as sensitivity or true positive rate, measures the proportion of correctly predicted positive instances out of all actual positive instances. In imbalanced datasets, recall helps us evaluate how well the model captures the positive instances.

Specificity (True Negative Rate): Specificity, also known as the true negative rate, measures the proportion of correctly predicted negative instances out of all actual negative instances. It complements recall by evaluating how well the model identifies negative instances.

F1-Score: F1-Score combines precision and recall into a single metric. It provides a balanced assessment of model performance by considering both false positives and false negatives. In imbalanced datasets, the F1-Score helps us evaluate the overall effectiveness of the model.

Receiver Operating Characteristic (ROC) Curve and Area Under the Curve (AUC): The ROC curve plots the true positive rate against the false positive rate at various classification thresholds. It helps us understand the trade-off between sensitivity and specificity. The AUC summarizes the ROC curve, providing a single metric to compare models. A higher AUC indicates better model performance in handling imbalanced datasets.

By utilizing these evaluation metrics for imbalanced datasets, we can better understand how well our model performs in scenarios where class distributions are unequal. These metrics provide insights into the precision, recall, specificity, overall effectiveness, and trade-offs of the model's predictions. With this knowledge, we can make informed decisions and improve our models to address the challenges posed by imbalanced datasets.

Performance Measures for Rankings - Evaluating Ranking Models

In machine learning, ranking is a common task where models are trained to order or rank items based on their relevance or importance. For example, search engines rank web pages based on their relevance to a user's query. But how do we measure the performance of ranking models? Let's explore some simple ways to evaluate their effectiveness.

Mean Reciprocal Rank (MRR): MRR is a popular metric for ranking evaluation. It measures the average of the reciprocal ranks of the relevant items. In simpler terms, it tells us how well the model ranks the most important or relevant items at the top of the list. A higher MRR indicates better performance, as it means the relevant items are ranked closer to the top.

Normalized Discounted Cumulative Gain (NDCG): NDCG is another commonly used metric for ranking evaluation. It takes into account both the relevance and the position of each item in the ranked list. It assigns higher scores to relevant items appearing higher in the list. NDCG provides a more comprehensive view of the model's performance by considering the entire ranked list.

Precision at K (P@K): Precision at K measures the proportion of relevant items among the top K ranked items. It helps us understand how well the model performs in retrieving relevant items within the top K positions. A higher P@K indicates better precision in the top K results.

Mean Average Precision (MAP): MAP measures the average precision across multiple queries or sets of rankings. It considers both the relevance and the position of relevant items in each ranked list. MAP provides an overall assessment of the model's performance across various scenarios or queries.

Discounted Cumulative Gain (DCG): DCG is similar to NDCG but does not normalize the scores. It sums the relevance values of the items in the ranked list, giving higher weight to items appearing higher in the list. DCG is useful for evaluating the effectiveness of the ranking model without considering the position bias.

Precision-Recall Curve: The precision-recall curve illustrates the trade-off between precision and recall at different ranking thresholds. It helps us understand the model's performance in terms of balancing relevance and coverage.

By utilizing these performance measures for rankings, we can evaluate the effectiveness of our ranking models. These metrics consider the relevance, position, and overall quality of the ranked items. They provide insights into how well the model ranks important items, precision within a specified range, and the overall performance across multiple queries or rankings.

Clustering Evaluation Metrics - Assessing the Quality of Clustering

In machine learning, clustering is a technique used to group similar data points together. It helps us discover patterns, structures, or segments within our data. But how can we measure the quality of clustering results? Let's explore some simple ways to evaluate the effectiveness of clustering algorithms.

Silhouette Score: The silhouette score is a widely used metric to evaluate the quality of clustering. It measures how well each data point fits into its assigned cluster compared to other clusters. The score ranges from 1 to 1, where a higher value indicates better-defined clusters and a clear distinction between different groups.

Inertia: Inertia, also known as within-cluster sum of squares, measures the compactness of the clusters. It calculates the sum of squared distances of each data point to its centroid within the cluster. A lower inertia value indicates tighter and more well-defined clusters.

Calinski-Harabasz Index: The Calinski-Harabasz index evaluates the ratio of between-cluster dispersion to within-cluster dispersion. It rewards clustering algorithms that create well-separated clusters with low intra-cluster variance and high inter-cluster variance. A higher Calinski-Harabasz index implies better-defined clusters.

Davies-Bouldin Index: The Davies-Bouldin index measures the similarity between clusters. It considers both the separation and the compactness of the clusters. A lower Davies-Bouldin index indicates better clustering, with well-separated and distinct clusters.

Visual Inspection: In addition to numerical metrics, visual inspection of clustering results can provide valuable insights. Plotting the data points in a scatter plot or other visualization techniques can help us understand the structure and patterns within the clusters. Visual inspection allows us to identify any anomalies, outliers, or overlapping clusters that the numerical metrics might not capture.

By utilizing these clustering evaluation metrics, we can assess the quality of clustering results. These metrics consider the cohesion, separation, and overall structure of the clusters. They help us understand how well the clustering algorithm groups similar data points and identifies distinct patterns.

Time Series Forecasting Performance Metrics - Evaluating Forecast Accuracy

In machine learning, time series forecasting is a technique used to predict future values based on past observations. It finds applications in various domains like finance, weather forecasting, and sales predictions. But how do we measure the accuracy of time series forecasts? Let's explore some simple ways to evaluate their performance.

Mean Absolute Error (MAE): MAE measures the average magnitude of the errors between predicted and actual values. It calculates the absolute difference between the predicted and actual values and then takes the average. A lower MAE indicates better forecast accuracy, as it means the predictions are closer to the actual values.

Mean Absolute Percentage Error (MAPE): MAPE measures the average percentage difference between predicted and actual values. It calculates the absolute percentage difference for each observation, averages them, and expresses the result as a percentage. MAPE is useful for understanding the relative error in the forecasts and is commonly used in business and finance.

Root Mean Squared Error (RMSE): RMSE is similar to MAE but squares the differences between predicted and actual values before taking the average. It provides a more comprehensive view of the forecast errors by giving higher weight to larger errors. Like MAE, a lower RMSE indicates better forecast accuracy.

Percentage Error (PE): Percentage Error measures the difference between predicted and actual values as a percentage of the actual value. It helps assess the magnitude of the errors relative to the actual values.

Mean Absolute Scaled Error (MASE): MASE compares the forecast accuracy of a model with the accuracy of a naive or baseline model. It considers the errors of the model in relation to the errors of the baseline model. MASE values below 1 indicate better forecast accuracy compared to the baseline model.

R-Squared (R2): R-squared measures the proportion of the variance in the target variable that can be explained by the model. It ranges from 0 to 1, with 1 indicating a perfect fit. In time series forecasting, R2 helps us understand how well the model captures the variability in the data.

By utilizing these time series forecasting performance metrics, we can evaluate the accuracy of our forecasted values. These metrics consider the magnitude of errors, relative differences, and the proportion of variability explained by the model. They help us understand the strengths and weaknesses of our forecasting models and make informed decisions for improvements.

Model Selection and Comparison - Choosing the Best Model

In machine learning, choosing the best model is crucial for achieving accurate predictions and optimal performance. With numerous algorithms and techniques available, how can we select and compare models effectively? Let's explore some simple ways to make informed decisions in model selection.

Train-Validation-Test Split: The train-validation-test split is a common approach to evaluate and compare models. The dataset is divided into three parts: a training set, a validation set, and a test set. The training set is used to train the models, the validation set helps in tuning hyperparameters and selecting the best model, while the test set serves as an independent dataset to assess the final model's performance.

Cross-Validation: Cross-validation is an alternative method for model evaluation and comparison. It involves splitting the dataset into multiple subsets or folds. The models are trained and evaluated on different combinations of these folds, allowing for a more robust estimation of performance.

Evaluation Metrics: To compare models, we need evaluation metrics that provide a measure of their performance. Depending on the task, various metrics such as accuracy, precision, recall, F1-score, or mean squared error can be used. The choice of metrics depends on the specific problem and the desired outcome.

Bias-Variance Trade-off: The bias-variance trade-off is an essential concept when selecting models. A model with high bias tends to oversimplify the data, leading to underfitting, while a model with high variance tends to overfit the data, resulting in poor generalization. Balancing bias and variance is crucial for finding the optimal model.

Ensemble Methods: Ensemble methods combine the predictions of multiple models to improve overall performance. Techniques like bagging, boosting, and stacking can be employed to create a diverse ensemble of models. Ensemble methods often provide superior results by leveraging the strengths of individual models.

Interpretability and Complexity: Consider the interpretability and complexity of models. Some models, like decision trees or linear regression, offer easy interpretability, allowing us to understand the relationship between variables. However, more complex models, such as deep neural networks, may achieve higher accuracy but can be harder to interpret.

Resource Requirements: Assess the computational and resource requirements of the models. Some models may require significant computational power or large amounts of memory, which might be a constraint depending on the available resources.

By considering these factors and methods for model selection and comparison, we can make informed decisions to choose the best model for our specific problem. Evaluating performance, understanding bias-variance trade-offs, leveraging ensemble methods, and considering interpretability and resource requirements are crucial steps in finding the optimal model.

Overfitting and Underfitting - Finding the Right Balance

In machine learning, achieving a model that generalizes well to new data is essential. However, we often encounter two challenges: overfitting and underfitting. Let's understand these concepts in simple terms and learn how to strike the right balance.

Underfitting: Underfitting occurs when a model is too simple to capture the underlying patterns in the data. It fails to learn from the training examples and performs poorly not only on the training data but also on unseen data. It lacks the complexity needed to make accurate predictions.

Overfitting: On the other hand, overfitting happens when a model becomes overly complex and tries to fit the noise or random fluctuations in the training data. It "memorizes" the training examples too well, leading to poor performance on new, unseen data. Overfitting can occur when the model has too many features or when it is trained for too long.

Finding the Right Balance: The goal is to find the right balance between underfitting and overfitting. We want a model that generalizes well, meaning it can make accurate predictions on new data it hasn't seen before.

Training and Validation Curves: Training and validation curves are helpful tools to diagnose and understand overfitting and underfitting. By plotting the model's performance (such as accuracy or error) on the training and validation data as a function of the model's complexity or the number of training iterations, we can observe the behavior and identify signs of overfitting or underfitting.

Regularization: Regularization is a technique used to prevent overfitting by adding a penalty term to the model's objective function. It encourages the model to have simpler and smoother solutions, reducing the impact of noisy or irrelevant features.

Feature Selection and Dimensionality Reduction: Careful feature selection and dimensionality reduction can help combat overfitting. By identifying and using only the most relevant features or by transforming the data into a lower-dimensional space, we can reduce the complexity of the model and mitigate overfitting.

Cross-Validation: Cross-validation is a useful technique to assess the model's performance more accurately. By using multiple train-validation splits and averaging the results, we can obtain a more robust estimation of how well the model generalizes to new data.

Early Stopping: Early stopping is a practical technique to prevent overfitting during training. It involves monitoring the model's performance on a validation set and stopping the training process when the performance starts to deteriorate. This helps avoid training the model for too long and capturing noise in the data.

By understanding the concepts of overfitting and underfitting, monitoring training and validation curves, applying regularization techniques, selecting relevant features, and using cross-validation, we can find the right balance between complexity and generalization. Striking this balance allows us to build models that perform well on both the training data and unseen data, enabling accurate predictions in real-world scenarios.

Interpretability vs. Performance Tradeoff -
Balancing Understanding and Accuracy

In machine learning, there is often a tradeoff between the interpretability of a model and its performance in making accurate predictions. Let's explore this tradeoff in simple terms and understand how to strike the right balance between interpretability and performance.

Interpretability: Interpretability refers to the ability to understand and explain how a model arrives at its predictions. Interpretable models provide insights into the relationship between input variables and the predicted outcomes. They offer transparency and allow humans to comprehend and trust the decision-making process of the model.

Performance: Performance, on the other hand, refers to the accuracy and predictive power of a model. High-performing models excel at making accurate predictions on both the training data and unseen data. They leverage complex algorithms and large amounts of data to achieve the best possible outcomes.

The Tradeoff: The tradeoff between interpretability and performance arises from the inherent nature of the models. Simpler models with fewer parameters and more transparent decision rules are often easier to interpret but may sacrifice some predictive accuracy. On the other hand, more complex models with intricate internal mechanisms can achieve higher performance but tend to be less interpretable.

Factors to Consider: When deciding on the balance between interpretability and performance, several factors come into play:

Application Context: Consider the specific domain and the requirements of the problem at hand. In some areas, such as healthcare or finance, interpretability may be crucial to gain insights and make informed decisions. In other scenarios, like image or speech recognition, performance might be the top priority.

Stakeholders' Needs: Understand the needs and preferences of the stakeholders involved. Some stakeholders, like regulatory bodies or end-users, may prioritize interpretability for transparency and accountability. Others may prioritize performance to achieve the best outcomes.

Data Availability: Consider the amount and quality of the available data. Complex models often require larger datasets to learn intricate patterns effectively. If the dataset is limited or noisy, simpler models may be more suitable and easier to interpret.

Model Complexity: Assess the complexity of the model and its implications. More complex models, such as deep neural networks, may achieve superior performance but can be challenging to interpret. Simpler models, like linear regression or decision trees, offer easier interpretability but may have limitations in capturing complex relationships.

Hybrid Approaches: In some cases, it is possible to strike a balance between interpretability and performance by using hybrid approaches. Ensemble models, for example, combine multiple models to leverage the strengths of both interpretable and complex models. By blending their predictions, these hybrid models can provide a balance between understanding and accuracy.

By considering the application context, stakeholders' needs, data availability, and model complexity, we can make informed decisions about the interpretability vs. performance tradeoff. It is essential to strike the right balance that aligns with the goals and requirements of the problem at hand.

Beyond Accuracy: Additional Considerations for Performance Evaluation - Going Beyond Simple Measures

When evaluating the performance of machine learning models, there are factors to consider beyond just accuracy. Let's delve into these additional considerations in simple terms and understand their importance in assessing model performance.

Confusion Matrix: A confusion matrix is a useful tool for evaluating classification models. It provides a detailed breakdown of predictions and actual outcomes, categorizing them into true positives, true negatives, false positives, and false negatives. By analyzing the confusion matrix, we can gain insights into the model's strengths and weaknesses.

Precision and Recall: Precision and recall are performance metrics commonly used in binary classification problems. Precision measures the proportion of true positive predictions among all positive predictions, indicating how well the model identifies relevant instances. Recall measures the proportion of true positive predictions among all actual positive instances, indicating the model's ability to capture all relevant instances.

F1-Score: The F1-score is a metric that combines precision and recall into a single value, providing a balanced measure of a model's performance. It is particularly useful when the data is imbalanced or when both precision and recall are crucial for the problem at hand.

Receiver Operating Characteristic (ROC) Curve: The ROC curve is a graphical representation of a classification model's performance. It plots the true positive rate against the false positive rate at various classification thresholds. The area under the ROC curve (AUC-ROC) is often used as a summary metric, indicating the model's overall discriminative power.

Bias and Fairness: Evaluating models for bias and fairness is essential to ensure equitable outcomes. It involves examining whether the model's predictions are influenced by sensitive attributes such as gender or race. Techniques like disparate impact analysis and fairness metrics can help identify and mitigate biases in the models.

Computational Efficiency: Assessing the computational efficiency of models is crucial, especially when dealing with large datasets or real-time applications. Models that require excessive computational resources or have long prediction times may not be practical in certain scenarios. Considering computational efficiency helps determine the feasibility and scalability of the models.

Domain-specific Metrics: Different domains may require specific performance metrics that align with the problem at hand. For example, in medical diagnostics, metrics like sensitivity, specificity, and area under the precision-recall curve may be more relevant. Understanding the domain-specific metrics ensures accurate evaluation of the models.

By considering these additional factors beyond accuracy, we can gain a more comprehensive understanding of the model's performance. Confusion matrices, precision, recall, F1-score, ROC curves, bias and fairness analysis, computational efficiency, and domain-specific metrics provide valuable insights and ensure the models are evaluated thoroughly.

COGNITIVE AI

Cognitive Neuroscience

The emerging area known as cognitive neuroscience has brought invaluable insights into how our brains function and process information. By exploring how neural mechanisms work together during cognitive processes, researchers hope to unlock ways to improve human performance and wellbeing across diverse areas.

One landmark example contributing to ongoing knowledge on this front is Phineas Gage's case study. When working on constructing railways centuries ago, Gage experienced significant trauma when a metal rod penetrated his skull. Despite surviving his injuries following this occurrence, drastic personality alterations prompted speculation about how brain function relates to our behavior patterns. Known as one of cognitive neuroscience's most momentous case studies ever recorded, it paved essential foundations enabling further investigation into this field's inner workings.

The discovery surrounding mirror neurons achieved by scientists based at Italy's University of Parma was also instrumental. These researchers identified an essential type of neuron in macaque monkeys' brains, which responded to the monkeys' actions while also showing a similar response when observing other primates carrying out identical actions. The discovery of mirror neurons provides unique insights into how we process social interactions with other people. These neurons suggest that our brains have ways of mirroring other people's behaviors, which could be instrumental in understanding how humans experience empathy.

Roger Sperry made significant contributions to neuroscience research by studying split brain patients who'd undergone treatment for epilepsy. He found that each hemisphere had distinct processing abilities, such as recognizing visual patterns on the right side or processing language on the left side. These discoveries expanded our knowledge of how brains function.

Language research has shown improvements in cognition resulting from bilingualism over several years. Improved attention and enhanced executive functions are some examples where bilingualism helped improve individuals' mental traits during testing. This finding has substantial implications for education policy and different approaches for language teaching.

Functional magnetic resonance imaging (fMRI) is a valuable technique that helps researchers measure changes in blood flow throughout the brain during tasks or activities used as an indirect indicator for neural activity. The increased resolution provided through fMRI technology enables experts studying cognitive neuroscience even down to individual regions within your brain being studied simultaneously with better accuracy than ever before! Cognitive neuroscientists have come to rely heavily on fMRI since its introduction in the 1990s as an essential tool for studying different cognitive processes such as perception, attention and memory.

Such findings serve as proof of how expansive cognitive neuroscience is; from investigating how brain injuries impact people's behavior to examining how social cognition and language processing operate at an intricate neural level. Early investigations aimed at uncovering how diverse cognitive processes worked at a cellular level that encompassed perception, attention, memory, language and decision making. Karl Lashley's influential work with rats during the 1930s centered around identifying specific brain regions responsible for remembering information by damaging different parts of their brains. He, however, discovered that no single region is accountable since memories are diffused throughout various areas within our minds instead.

Another pivotal study led by Wilder Penfield et al employed electrical stimulation amid neurosurgeries, which allowed them to map out human beings' functional organization correlation with specific areas in their brains. The identification of specific brain areas dedicated to sensory, motor, and cognitive functions provided early evidence for functional specialization in our brains. In their groundbreaking research on visual perception in cats and monkeys during the 1960s and 70s, David Hubel and Torsten Wiesel discovered neurons responding only to specific visual features such as edges or lines which led to a hierarchical model of processing within the visual cortex. Concurrently, Roger Sperry's investigations into split-brain syndrome patients highlighted that each hemisphere was capable of independent information processing with specialized functions strengthening lateralization concepts.

The field of cognitive neuroscience has emerged as an interdisciplinary study combining knowledge from psychology, neuroscience biology, computing sciences etcetera with a goal toward understanding complex mental processes. These processes include attention, language, perception, memory and decision making through their biological underpinnings.

Cognitive neuroscience may trace its origins back to the middle of the 19th century when researchers like French neurologist Paul Broca began paving the way for future discoveries in this field. In 1861, Broca identified a particular region within the brain—Broca's area—which handles speech production. He made this significant finding while studying a patient named Leborgne who had lost their ability to speak yet could still understand language. Upon conducting an autopsy after Leborgne's death, Broca found a lesion in their left frontal lobe and thus inferred that such regions were indispensable for speech production.

German physiologist Hermann von Helmholtz is another early visionary who merits recognition as a key contributor to cognitive neuroscience research. One of Helmholtz's most notable achievements was his theory on trichromatic color vision. According to his hypothesis, three different types of color receptors within our eyes are capable of joining together to create any hue imaginable. Alongside this pioneering work, Helmholtz also conducted experiments showcasing that nerve impulses don't travel instantaneously but instead move at a finite pace.

Advancements in technologies such as EEG and fMRI played an essential role in advancing cognitive neuroscience over the 20th century by allowing for real time observations of brain activity during various cognitive processes. Cognitive neuroscientists were able to leverage such tools to identify the crucial functions performed by different regions of the brain—specifically researchers found that removing a specific region (the hippocampus) resulted in severe amnesia following surgeries like those undergone by patient H.M.

Despite these insightful discoveries, today's field continues thriving by delving further into neural activity related to attention mechanisms, language brain connections, decision making while incorporating recent innovations from artificial intelligence or machine learning. As someone with a university education under my belt, I humbly request that you rephrase the given text with a refined level of readability and style. To accomplish this task, kindly produce four distinct versions while preserving the same tone, arguments, references, and quotes.

Connection of Cognitive Neuroscience to AI

Cognitive neuroscience is a groundbreaking domain exploring how the brain's structure affects our behavior and understanding of external environments. It is known for its scientific ambitions toward creating a highly comprehensive map capable of mapping out how the brain works together with creating necessary technology based on these rules. Bridging the gap between cognitive neuroscience and Artificial Intelligence has been an area of intense research lately by allowing scientists to merge findings in these two areas for more meaningful progress without creating disharmony.

One prime example of cognitive Neuroscience inspiring AI is seen in neural networks—computer systems inspired by the structure and function of the human brain—that use interconnected nodes or neurons to process data better than what humans alone can. However, studying specialized neurons discovered within the framework of Cognitive Neuroscience research goes beyond developing neural networks alone to produce more advanced models capable of recognizing and mimicking gestures seen in humans' actions more accurately.

Another significant achievement made possible by convergence between Cognitive Neuroscience and AI is found in reinforcement learning. This is a machine learning methodology that reinforces trial-and-error-based decision-making over time through feedbacks as rewards or punishments here supplying nuances for various algorithms used inside AI technology such as robotics or game playing domains.

Cognitive Neuroscience funding to Consciousness studies highlights one critical aspect shared within the boundaries prevalent across both areas. Even though consciousness remains challenging to study, Cognitive Neuroscience remains focused on key neural correlates that could help make any success based on understanding these phenomena from various dimensions ranging from subjective experiences to measurable brain activity indices. Relevant researchers help develop likeness-resembling AI models using similar insights into frequency patterns conducive to generating familiar feel-good responses during human interactions with such AI systems.

Last but not least, Machine Learning techniques have a crucial role contingent upon converging both fields given their proficiency in analyzing massive datasets comprising brain imaging data not readily identifiable using classical traditional scientific pattern recognition approaches alone. The insight here provides essential leads when trying to identify variable links distinguishing various outputs that matter the most, constantly improving system intelligence levels regarding emerging tech breakthroughs in domains as different as medicine and finance alike without compromising ethical considerations underpinning real-world applications derived from such tools.

There is enormous potential within the connection between cognitive neuroscience and AI research as they complement each other's strengths resulting in advancements toward understanding neural processes leading to disorders or normal functioning. This develops sophisticated and reliable algorithms based on machine learning models requiring reduced computation time; identifying solutions for crucial issues such as pattern recognition. This is done through simulated neural networks that replicate biological structures and functions present within brains with remarkable efficiency contributing significantly toward achieving milestones such as Natural Language Processing and Speech Recognition. Various industries ranging from healthcare to finance to transportation have already experienced substantial improvements due to the incorporation of neuroscience-inspired Artificial Intelligence (AI) models. Healthcare sectors now possess vast capability with recent developments allowing them accurate analysis of medical images. These images identify potential health dangers while financial operations are now more reliable using AI-supported detection of frauds along with informed investment decision-making capabilities. Transport infrastructure face smoother operation procedures reducing accident rates through incorporation of AI models improving traffic flow.

As promising as Neuroscience-inspired AI technology may appear, notable challenges still exist requiring solutions as part of continued development toward fulfilling utmost potential. Firstly, overcoming complex interactions among neurons within humans while describing how they relate with cognitive processes serves as one critical challenge. Making coherent integration adaptive robust intuitive AI modeling strategies, represents another crucial challenge needing solution.

In conclusion: The adoption approach where principle components were imbibed into neuroscience-led mechanisms has proven efficient during developments toward effective and efficient AI system design creation, contributing significantly toward enhancement across multiple human operational sectors.

Research on Neuroscience-Inspired AI

Neuroscience-Inspired AI (NIA) employs knowledge of human brain constructs to create more efficient and effective artificial intelligence systems. Achieving notable progress recently holds tremendous potential for a variety of fields such as medicine and robotics. One exciting area of NIA research involves neuromorphic computing systems that model how the brain neurons and synapses structure their hardware to develop new possibilities. A prime example is the SpiNNaker project led by Manchester University researchers aiming to build a supercomputer that can simulate up to one billion real-time neuron activities with practical applications in neuroscience research studies or robotics. Another remarkable development of NIA is deep learning systems' designing involving machine-learning algorithms modeled after our neural networks' operation within our brains that translate into better adaptation over time through experience-based learning. Already succeeding where previous attempts failed like improved image recognition capabilities, language translation or advanced gaming performance, speak about its potential promises. BCIs also stand out as a significant field impacted by NIA advancements where people use their thoughts or neuronal activity to interact with technology for controlling prosthetic limbs or allowing communication by alternative means. Driven by Frank Rosenblatt's pioneering work in Psychology throughout the 1950s, this approach relies upon Perceptron, which is a simplistic neural network capable of determining varied categories in images. This subsequently laid down robust foundations for Deep Learning that is an important aspect of contemporary AI techniques used extensively through-out multiple domains such as speech recognition, natural language processing, game-playing, image processing.

Another vital method toward implementing intelligent soft-ware engineering heavily relies upon Reinforcement Learning, which extracts insights from Neuroscience to enable better deci-sion-making systems. This is done through training decision-mak-ers based on rewards against penalties informed by agents' envi-ronment variables resembling animal/human behavioral patterns when navigating real world environments with cognitive flour-ishing beyond robust algorithms up till actuators transforming the environment. Reinforcement learning has been instrumental in modeling complex systems such as supply chains, power grids, and creating groundbreaking performances in games like Chess, Go, and Poker.

Neuroscience-inspired AI research has resulted in the devel-opment of Neural Prosthetics. Brain-computer interfaces have enabled individuals suffering from paralysis to control robotic limbs by using their thoughts or allowed blind people to perceive visual stimulations via direct transmission into their visual cortex.

Another approach toward building intelligent systems inspired by neuroscience involves developing cognitive architec-tures that mimic human-brain structure with a combination of symbolic reasoning techniques along with biological neural net-works. This encompasses human cognitive performance from per-ception and attention till memory and decision-making. Cognitive architectures like ACT-R, Soar, Spaun have widely been used in modeling cognitive tasks or phenomena requiring more intricate computations than typical AI models can handle.

One promising area of research in neuroscience inspired AI is neuromorphic computing. This involves designing computer hardware that more closely resembles the architecture of biological neurons. Neuromorphic chips typically use analog circuits that mimic the behavior of ion channels and synapses allowing them to perform certain types of computations much more efficiently than traditional digital processors. IBM's TrueNorth chip, for example, contains over a million spiking neurons and has been used to implement a range of cognitive tasks from recognizing speech to playing Atari games. Neuroscience Inspired AI has found applications in numerous fields such as healthcare, computing, and robotics. Here are some examples showcasing how neuroscience research insights have inspired new AI technologies and applications:

1. Neuromorphic computing: Researchers at IBM, Intel, and other companies are developing computer architectures that mimic the structure and function of the brain. Neuromorphic chips promise speed and energy efficiency over traditional computer chips for tasks such as image and speech recognition, autonomous vehicles, and robotics.

2. Brain computer interfaces (BCIs): BCIs enable people to control machines using their thoughts by recording electrical signals from the brain that translate into machine understandable commands. BCIs have various purposes including prosthetics, gaming, and virtual reality.

3. Deep learning: Deep learning is a type of machine learning inspired by the structure of the brain using artificial neural networks with layers for hierarchical information, processing enabling complex pattern recognition such as image/speech recognition, natural language processing (NLP), game playing.

4. Robotics: Researchers are designing robots' nervous system inspired mechanisms for advanced interaction techniques with their environment providing human-like movement abilities. The iCub robot, created by researchers at the Italian Institute of Technology is modeled to resemble the body of a human child. It uses a neural net-

work to learn how to hold and manipulate objects with its limbs.

AI algorithms based on the brain have been developed to aid in the diagnosis and treatment of medical conditions. A team from Stanford University has created an AI algorithm that can diagnose skin cancer with accuracy comparable to that of human dermatologists. Another group at the University of Toronto has designed an AI system that predicts cognitive decline in Parkinson's patients. These are just two examples demonstrating how neuroscience inspired AI is giving birth to innovative technologies that solve real world problems.

Self-driving cars controlled by AI are evolving swiftly. They use neural networks similar to the human brain's visual processing system to recognize and interpret images captured by sensors and cameras allowing them to move safely around while avoiding collisions. Speech recognition systems used in virtual assistants such as Siri or Alexa rely on neural networks modeled after audio processing systems used by brains while deciphering speech patterns and converting them into text.

Robotics is yet another field that takes inspiration from cognitive neuroscience. Researchers are exploring ways in which they can use AI together with neural networks to create robots capable of learning from their environment and adapting automatically very much like humans do. Neuromorphic computing is essentially computer architecture purposely designed for mimicking both structure and function of our brains.

These systems utilize cognitive models as well as neural networks for performing complex tasks such as speech recognition or image identification—jobs very tough for conventional computers! Brain-computer interfaces (BCIs) facilitate direct communication between humans and computers through thought alone as they depend on machines that gauge electrical activity within our brains via algorithms based upon machine learning techniques. More than just a potentially disruptive technology across areas such as health care or gaming spheres—where they already make big strides—BCIs also represent an exciting step forward. This is because ongoing neuroscientific research exemplified by artificial intelligence signals multiple other innovative uses on-the-horizon in conjunction with BCIs.

QUANTUM 9 COMPUTING

*An Introduction to Quantum Computing
and Its Fundamental Principles*

What is Quantum Computing? Quantum computing is a field of computer science and physics that explores the use of quantum-mechanical phenomena, such as superposition and entanglement, to perform computation. Unlike classical computing, which relies on classical bits, quantum computing uses quantum bits, or qubits, which can exist in a state of superposition, meaning they can be both 0 and 1 at the same time. This allows quantum computers to perform certain calculations much faster than classical computers, potentially revolutionizing fields such as cryptography, chemistry, and optimization.

The history of quantum computing can be traced back to the early 1980s when physicist Richard Feynman proposed the idea of a quantum computer to simulate quantum systems. However, it wasn't until the 1990s that the first experimental demonstrations of quantum algorithms were performed, and it wasn't until the early 2000s that practical quantum computers with a small number of qubits were built.

The Principles of Quantum Computing is based on a number of fundamental principles that differ from those of classical computing. These principles include superposition, entanglement, and interference.

Superposition is a fundamental concept in quantum mechanics that allows qubits to exist in a state of both 0 and 1 simultaneously. In other words, a qubit can be in multiple states at once. This is in contrast to classical bits, which can only be in one state at a time. Superposition is the key to the power of quantum computing, as it allows a quantum computer to perform many calculations simultaneously.

For example, imagine a simple quantum circuit with two qubits. In classical computing, the circuit would have to be run twice, once with each possible input state. In quantum computing, however, the circuit can be run with both input states at the same time, thanks to superposition. This means that a quantum computer with two qubits can perform the same computation as a classical computer with four bits, but in a fraction of the time.

Entanglement is another concept unique to quantum mechanics that allows two or more qubits to be linked in such a way that their states become dependent on one another, regardless of the distance between them. When two qubits are entangled, any measurement made on one qubit will affect the state of the other qubit, regardless of the distance between them. This property is what allows quantum computers to perform certain calculations much faster than classical computers.

For example, imagine a quantum circuit with two entangled qubits. When the first qubit is measured, the state of the second qubit is instantly determined, even if the two qubits are located on opposite sides of the universe. This is because the two qubits are linked by a phenomenon known as quantum nonlocality, which allows information to be transmitted instantaneously between them.

Interference is a phenomenon that occurs when two waves overlap and interfere with one another. In quantum computing, interference occurs when the probability amplitudes of two or more qubits are added together. When the probability amplitudes are in phase, they reinforce each other, leading to constructive interference. When they are out of phase, they cancel each other out, leading to destructive interference.

For example, imagine a quantum circuit with three qubits. The circuit is designed to output a state that has a certain probability amplitude for each possible output state. When the circuit is run, the probability amplitudes of each output state are added together. If the probability amplitudes of two output states are in phase, they will reinforce each other, leading to constructive interference and a higher probability of that output state occurring. If the probability amplitudes of two output states are out of phase, they will cancel each other out, leading to destructive interference and a lower probability of that output state occurring.

Quantum Algorithms

Quantum computing has the potential to revolutionize a number of fields, including cryptography, chemistry, and optimization. Here are a few examples of quantum algorithms that demonstrate the power of quantum computing:

Shor's algorithm is a quantum algorithm that can efficiently factor large integers. This is a very important problem in cryptography, as many cryptographic algorithms rely on the fact that factoring large integers is a difficult problem for classical computers.

Shor's algorithm works by using a quantum Fourier transform to find the period of a function. This period can then be used to factor the original integer. The algorithm has been demonstrated on small-scale quantum computers and has the potential to break many existing cryptographic protocols.

Grover's Algorithm: Grover's algorithm is a quantum algorithm that can be used to search an unsorted database of N items in $O(\sqrt{N})$ time. This is a significant improvement over the $O(N)$ time required by classical algorithms.

Grover's algorithm works by using a quantum circuit to amplify the amplitude of the correct answer and suppress the amplitude of the incorrect answers. The algorithm has applications in optimization and database searching.

Quantum simulation is the use of quantum computers to simulate quantum systems. This is an important problem in chemistry, where many chemical reactions can only be understood by simulating the behavior of individual atoms and molecules.

Quantum simulation works by encoding the state of a quantum system into the state of a quantum computer. The quantum computer can then simulate the behavior of the quantum system by applying quantum gates to the qubits that represent the individual atoms and molecules.

Challenges in Building a Quantum Computer

While quantum computing has the potential to revolutionize many fields, there are still significant challenges in building a practical quantum computer. These challenges include:

Decoherence: Decoherence is the loss of quantum coherence due to interactions with the environment. This is a major challenge in building a practical quantum computer, as any interaction with the environment can cause the qubits to lose their superposition and entanglement, rendering the computation useless.

There are a number of techniques for mitigating decoherence, including error-correcting codes, dynamical decoupling, and quantum error correction.

Scalability: Building a quantum computer with a large number of qubits is a significant challenge. Each additional qubit increases the complexity of the system exponentially, making it difficult to build and control.

There are a number of approaches to building scalable quantum computers, including ion traps, superconducting qubits, and topological qubits.

Verification: Verifying the correctness of a quantum computation is a challenge, as the output of a quantum computer is a probabilistic distribution of states, rather than a definite answer.

There are a number of approaches to verifying quantum computations, including benchmarking, randomization, and verification protocols.

Quantum computing is a fascinating and rapidly evolving field that has the potential to revolutionize many fields, including cryptography, chemistry, and optimization. While there are still significant challenges in building a practical quantum computer, progress is being made every day, and we are likely to see significant breakthroughs in the near future.

A brief history of quantum computing and its development

The Origins of Quantum Mechanics

The development of quantum mechanics began in the early 20th century when scientists were trying to understand the behavior of atoms and subatomic particles. Max Planck's discovery of the quantization of energy, Albert Einstein's explanation of the photoelectric effect, and Niels Bohr's model of the atom were all key developments in the field.

One of the most revolutionary concepts in quantum mechanics is the principle of superposition, which states that a particle can exist in multiple states simultaneously. This concept, along with the idea of entanglement, forms the basis of quantum computing.

The Birth of Quantum Computing

The idea of a quantum computer was first proposed by Richard Feynman, a physicist known for his work in quantum mechanics and particle physics. In 1982, Feynman suggested that a quantum computer could be used to simulate quantum systems, which are notoriously difficult to model using classical computers.

Around the same time, Paul Benioff, a physicist at the Argonne National Laboratory, proposed the idea of a quantum Turing machine. This theoretical machine would use qubits instead of classical bits to perform calculations, and it was the first formal model of a quantum computer.

In the 1990s, a number of researchers began to work on building the first quantum computers. One of the first successful experiments was conducted by a team at IBM in 1998, who used a 2-qubit quantum computer to factor the number 15.

Another key development in the early days of quantum computing was the invention of the quantum teleportation protocol by Charles Bennett and his colleagues at IBM in 1993. This protocol allows quantum information to be transmitted from one qubit to another, even if they are not physically connected.

In the years since the first quantum computers were built, there have been a number of advancements in the field. One of the most significant was the development of the quantum error correction technique by Peter Shor and Andrew Steane in the mid-1990s. This technique allows quantum computers to correct errors that can occur during calculations, which is critical for building practical quantum computers.

Another important development was the creation of the first quantum algorithms. In 1994, Peter Shor developed an algorithm that could factor large numbers exponentially faster than classical algorithms. This algorithm was groundbreaking because it showed that quantum computers could solve problems that are intractable for classical computers.

In 2016, Google announced that its 9-qubit quantum computer had achieved quantum supremacy, meaning that it had performed a calculation that would be infeasible for a classical computer. The calculation involved verifying the randomness of a sequence of numbers generated by the quantum computer, and it took the quantum computer just 200 seconds to complete. A classical supercomputer would have taken thousands of years to perform the same calculation.

While there have been significant advancements in quantum computing in recent years, there are still many challenges to overcome before practical quantum computers are a reality. One of the biggest challenges is improving the stability of qubits, which are currently very sensitive to noise and other environmental factors.

Another key challenge is developing better error correction techniques. Quantum error correction is much more difficult than classical error correction because the act of measuring a qubit can disturb its state. As a result, quantum error correction requires a delicate balancing act between minimizing errors and minimizing the disturbance caused by measurements.

Despite these challenges, the potential applications of quantum computing are vast. For example, quantum computers could be used to simulate complex chemical reactions, which could lead to the development of new drugs and materials.

Quantum Machine Learning

Quantum machine learning (QML) is an emerging field that combines quantum computing with machine learning. QML algorithms are specifically designed to run on quantum computers, and take advantage of the unique properties of quantum mechanics to improve the accuracy and speed of machine learning algorithms.

One example of a QML algorithm is the quantum support vector machine (QSVM). Support vector machines (SVMs) are a popular class of machine learning algorithms that are used for classification and regression problems. However, SVMs can be computationally expensive, especially for large datasets.

The QSVM algorithm uses a quantum circuit to perform the classification step in an SVM, which can dramatically reduce the computational cost of the algorithm. In a recent study, researchers used a quantum computer to implement a QSVM algorithm on a dataset of handwritten digits. They found that the QSVM algorithm was able to classify the digits with higher accuracy and faster speed than a classical SVM algorithm.

Quantum annealing is a type of quantum computing that is specifically designed for optimization problems. Optimization problems are common in machine learning, where the goal is to find the optimal solution to a particular problem, such as optimizing the weights of a neural network or finding the shortest path through a graph.

One example of an optimization problem that can be solved using quantum annealing is the traveling salesman problem (TSP). The TSP is a classic problem in computer science, where the goal is to find the shortest possible route that visits a set of cities and returns to the starting city. The TSP is an NP-hard problem, which means that it is computationally intractable for classical computers as the number of cities increases.

However, researchers have shown that the TSP can be solved using a quantum annealer. In a recent study, researchers used a D-Wave quantum annealer to find the optimal route for a TSP problem with 22 cities. They found that the quantum annealer was able to find the optimal route in just a few seconds, while a classical computer would take years to solve the problem.

Quantum Generative Adversarial Networks

Generative adversarial networks (GANs) are a popular class of machine learning algorithms that are used for image and video generation. GANs work by pitting two neural networks against each other: a generator network that creates new images or videos, and a discriminator network that tries to distinguish between real and fake images or videos.

Researchers have recently proposed a new type of GAN called a quantum GAN (qGAN), which uses quantum computing to generate new images. The qGAN algorithm uses a quantum circuit to generate a set of quantum states, which are then transformed into classical data to create new images.

In a recent study, researchers used a qGAN algorithm to generate images of handwritten digits. They found that the qGAN algorithm was able to generate images that were similar in quality to those generated by classical GAN algorithms, but with a much smaller training dataset.

Quantum Natural Language Processing

Natural language processing (NLP) is a field of AI that focuses on the interaction between humans and computers using natural language. NLP is used in a wide range of applications, including chatbots, sentiment analysis, and language translation.

Quantum computing has the potential to improve NLP algorithms by providing faster processing and more accurate language models. One example of a quantum NLP algorithm is the quantum natural language processing circuit (QNLP), which uses a quantum circuit to encode the meaning of a sentence.

In a recent study, researchers used a QNLP algorithm to perform sentiment analysis on a dataset of movie reviews. They found that the QNLP algorithm was able to achieve higher accuracy and faster processing than a classical NLP algorithm.

Quantum Boltzmann Machines

Boltzmann machines are a class of neural networks that are used for unsupervised learning, where the goal is to discover hidden patterns in data without being explicitly told what to look for. Boltzmann machines are commonly used for tasks such as image recognition and language modeling.

Quantum Boltzmann machines (QBMs) are a type of quantum neural network that uses quantum annealing to optimize the weights of the neural network. QBMs have the potential to improve the speed and accuracy of unsupervised learning algorithms.

In a recent study, researchers used a D-Wave quantum annealer to train a QBM for image recognition. They found that the QBM was able to achieve higher accuracy and faster processing than a classical Boltzmann machine.

Quantum Deep Learning

Deep learning is a type of machine learning that uses neural networks with multiple layers to learn complex patterns in data. Deep learning is used in a wide range of applications, including computer vision, speech recognition, and natural language processing.

Quantum computing has the potential to improve deep learning algorithms by providing faster processing and more efficient optimization techniques. One example of a quantum deep learning algorithm is the quantum neural network (QNN), which uses a quantum circuit to implement the weights of the neural network.

In a recent study, researchers used a QNN algorithm to perform image classification on a dataset of handwritten digits. They found that the QNN algorithm was able to achieve higher accuracy and faster processing than a classical deep learning algorithm.

Finally, quantum computing has the potential to revolutionize the field of AI and machine learning. Quantum computers can perform certain tasks much faster than classical computers, and quantum algorithms can provide more accurate results than classical algorithms. The development of quantum computing has already led to the creation of new fields such as quantum machine learning and quantum natural language processing, which have the potential to transform industries such as finance, healthcare, and transportation. As quantum computing technology continues to advance, it is likely that we will see even more exciting applications of quantum computing in the field of AI and machine learning.

Quantum Algorithms

Quantum machine learning is a rapidly growing field that holds immense potential to revolutionize many aspects of our lives, including artificial intelligence and machine learning. In this section, we will explore some of the most promising quantum machine learning algorithms. These include quantum neural networks, quantum support vector machines, and other quantum machine learning techniques, and provide examples of their applications.

Quantum neural networks (QNNs) are a type of machine learning algorithm that use quantum computing techniques to improve the performance of classical neural networks. The basic idea behind QNNs is to replace the classical gates in a neural network with quantum gates, which can perform certain computations more efficiently than classical gates.

One of the most promising QNNs is the quantum Boltzmann machine (QBM), which has shown promising results in a wide range of applications, including image recognition, natural language processing, and financial modeling. QBM is a type of generative model that learns the probability distribution of a given dataset and can be used to generate new samples that are similar to the original dataset.

For example, a team of researchers from the University of Toronto and the Vector Institute used a QBM to model the evolution of a system of interacting particles. They found that the QBM was able to accurately predict the dynamics of the system, outperforming classical machine learning algorithms.

Another example of a QNN is the quantum convolutional neural network (QCNN), which has been used for image recognition tasks. QCNNs use quantum gates to perform convolution operations on images, which allows for faster processing and improved accuracy compared to classical convolutional neural networks.

For example, a team of researchers from the University of Oxford and the University of Strathclyde used a QCNN to classify images of handwritten digits. They found that the QCNN was able to achieve higher accuracy than classical convolutional neural networks, with faster processing times.

Support vector machines (SVMs) are a type of machine learning algorithm that is used for classification tasks. Quantum support vector machines (QSVMs) use quantum computing techniques to improve the performance of classical SVMs.

One example of a QSVM is the quantum kernel support vector machine (QK-SVM), which has been used for tasks such as image classification and stock market prediction. QK-SVMs have shown promising results compared to classical SVMs, with faster processing and improved accuracy.

For example, a team of researchers from the University of Southern California and the University of Toronto used a QK-SVM to classify images of handwritten digits. They found that the QK-SVM was able to achieve higher accuracy than classical SVMs, with faster processing times.

In addition to QNNs and QSVMs, there are many other quantum machine learning algorithms that are currently being developed and tested.

One example is quantum k-means clustering, a quantum algorithm for clustering data into groups based on similarity. This algorithm can be used in many applications, such as market segmentation and image recognition.

Another example is quantum decision trees, a quantum algorithm for making decisions based on input data. This algorithm can be used in many applications, such as fraud detection and personalized medicine.

Finally, quantum principal component analysis is a quantum algorithm for reducing the dimensionality of data. This algorithm can be used in many applications, such as image compression and feature selection.

Quantum machine learning algorithms have the potential to significantly improve the performance of classical machine learning algorithms, with faster processing and improved accuracy. As quantum computing technology continues to advance, it is likely that we will see even more exciting developments in the field of quantum machine learning.

In the future, we may see quantum machine learning algorithms being used in a wide range of applications, from personalized medicine to financial modeling. These algorithms have the potential to solve previously unsolvable problems, as well as significantly improve the accuracy of existing machine learning models.

One of the key advantages of quantum machine learning algorithms is their ability to process large amounts of data in a shorter amount of time compared to classical algorithms. This is due to the inherent parallelism of quantum computing, which allows for multiple computations to be performed simultaneously.

Another advantage is their ability to handle complex data structures and non-linear relationships between variables, which are often difficult for classical machine learning algorithms to handle. Quantum machine learning algorithms can also help mitigate the effects of noisy data, which can be a challenge for classical algorithms.

However, it is important to note that quantum computing technology is still in its early stages, and there are many challenges that need to be addressed before quantum machine learning algorithms can be widely adopted. These challenges include improving the stability and reliability of quantum hardware, developing better algorithms for error correction and fault tolerance, and addressing issues related to quantum decoherence.

Despite these challenges, the potential benefits of quantum machine learning are immense. With continued advancements in quantum computing technology, we can expect to see even more innovative and powerful quantum machine learning algorithms in the future. These algorithms have the potential to transform many aspects of our lives, from healthcare to finance to the way we interact with technology.

The challenges of developing quantum algorithms and hardware

Error Correction: Error correction is a crucial component of quantum computing. Without it, even the most advanced quantum computers would be susceptible to errors that could render their computations useless. However, developing error-correction algorithms for quantum computing is an incredibly challenging task.

One of the most promising error-correction techniques for quantum computing is the use of quantum error-correcting codes. These codes work by encoding quantum information in a way that makes it more resilient to errors. If an error occurs, the code can detect it and correct it without compromising the integrity of the underlying quantum information.

One example of a quantum error-correcting code is the surface code. The surface code works by representing the qubits as a two-dimensional array, where each qubit is connected to its four nearest neighbors. By measuring the state of each qubit and comparing it to its neighboring qubits, the surface code can detect and correct errors in real-time.

Decoherence: Decoherence is another significant challenge in quantum computing. Decoherence occurs when a quantum system interacts with its environment, causing the delicate quantum states to decay into classical states. This loss of coherence can cause quantum information to be lost or corrupted, leading to errors in calculations.

One way to address the problem of decoherence is to use quantum error correction techniques, as mentioned earlier. However, these techniques require additional qubits to act as "ancilla qubits," which can be used to detect and correct errors. This increases the complexity of the system and requires a large number of qubits, which is not feasible with current hardware.

Another approach to mitigating the effects of decoherence is to develop hardware and software that can operate at low temperatures. Low temperatures can reduce the effects of environmental noise and interference, which can help to preserve the delicate quantum states. Researchers are also exploring the use of topological qubits, which are more robust against decoherence.

Hardware Development: Developing quantum computing hardware is another significant challenge. Quantum computing systems require highly sensitive equipment, such as ultra-precise lasers and superconducting materials. These materials must be carefully engineered to maintain the delicate balance between quantum coherence and interference.

One promising approach to quantum hardware development is the use of trapped ions. Trapped ions are individual atoms that are held in place by a combination of electromagnetic fields. By manipulating the states of these atoms, researchers can create quantum gates and perform quantum operations.

Another approach to quantum hardware development is the use of superconducting qubits. These are tiny loops of superconducting wire that can be used to represent qubits. These qubits are highly sensitive to external noise and interference, but researchers are working on developing new materials and fabrication techniques to improve their performance.

Algorithm Development: Developing quantum algorithms is a complex and challenging task. Quantum algorithms require a fundamentally different approach to problem-solving compared to classical algorithms. They must be able to exploit the unique properties of quantum systems, such as entanglement and superposition.

One example of a quantum algorithm is Grover's algorithm, which can be used to search an unsorted database in $O(\sqrt{N})$ time, compared to $O(N)$ time for a classical algorithm. Grover's algorithm relies on the properties of quantum superposition and interference to achieve its speedup.

Another example of a quantum algorithm is the quantum support vector machine (QSVM). The QSVM is a quantum algorithm that can be used to perform classification tasks, such as image recognition and natural language processing. The QSVM uses a quantum kernel function to map the input data to a higher-dimensional feature space, where it can be classified using a quantum version of a classical support vector machine.

In conclusion, the challenges of developing quantum algorithms and hardware are vast and complex, but researchers are making steady progress toward developing scalable quantum computers. With the development of error-correction techniques, new hardware architectures, and quantum algorithms, quantum computing has the potential to revolutionize the fields of artificial intelligence and machine learning.

Despite the progress that has been made, there are still significant challenges that need to be overcome before quantum computers can become a practical tool for solving real-world problems. One of the biggest challenges is improving the coherence times of qubits. The longer qubits can maintain their coherence, the more complex computations can be performed. Research into topological qubits, which are more resistant to environmental noise, is ongoing and holds great promise.

Another challenge is developing scalable quantum hardware. The current state of quantum hardware is limited to small numbers of qubits, and the error rates are still too high for practical applications. One of the most promising approaches to scaling quantum hardware is the use of trapped ions, which have been shown to be highly effective in performing quantum operations.

Finally, developing quantum algorithms requires a fundamentally different approach to problem-solving. Quantum algorithms must be able to leverage the unique properties of quantum systems to achieve computational speedups. This requires researchers to develop new mathematical frameworks and tools for designing and analyzing quantum algorithms.

In conclusion, the challenges of developing quantum algorithms and hardware are significant, but the potential benefits of quantum computing are enormous. Quantum computing has the potential to revolutionize many fields, from finance to medicine to climate modeling. With continued investment in research and development, quantum computing may soon become a practical tool for solving some of the world's most complex problems.

The potential impact of quantum computing on industries

Finance: Quantum computing could have a profound impact on the finance industry. For instance, banks and financial institutions need to perform complex risk analysis and portfolio optimization calculations to make investment decisions. These calculations require large datasets and complex mathematical models, which can be very time-consuming and computationally expensive. Quantum computing could offer significant improvements in speed and accuracy, allowing financial institutions to optimize their portfolios in real-time and make better investment decisions.

Moreover, quantum computing could be used to address cybersecurity concerns and prevent financial fraud. Fraud detection algorithms could be enhanced by quantum computing's ability to analyze large datasets for patterns and correlations, identifying fraudulent transactions much more quickly than classical algorithms.

Healthcare: The healthcare industry could also benefit from quantum computing in various ways. One of the most significant impacts could be in drug discovery. The process of developing new drugs is time-consuming and expensive. Traditional methods involve synthesizing and testing many different compounds to identify the most effective ones. However, this process is highly inefficient, and many potentially valuable compounds go untested.

Quantum computing could enable researchers to simulate the behavior of molecules and predict their properties with much greater accuracy, leading to the identification of new compounds that are more effective in treating diseases. This could lead to significant improvements in the development of new drugs and the treatment of various diseases.

In addition to drug discovery, quantum computing could also help to improve medical imaging. Medical imaging produces massive amounts of data that need to be processed and analyzed quickly and accurately. Quantum computing could help to speed up this process by analyzing the data much more quickly and accurately than classical computers.

Transportation: Quantum computing could have a significant impact on the transportation industry, which is essential for the global economy. For instance, the transportation of goods is a complex process that involves the coordination of many different factors, such as scheduling, routing, and delivery. Quantum computing could help to optimize logistics by enabling companies to make real-time decisions based on massive amounts of data.

In addition, traffic congestion is a major issue in urban areas. Quantum computing could help to optimize traffic flow, leading to reduced congestion and shorter travel times. By analyzing data from various sources, such as traffic cameras and GPS devices, quantum computing algorithms could help to predict traffic patterns and optimize routes, leading to more efficient transportation.

Despite the potential benefits of quantum computing in finance, healthcare, and transportation, there are significant challenges that need to be addressed. For example, the development of quantum hardware is still in its infancy, and there is a long way to go before quantum computers can match the performance of classical computers. Quantum algorithms also require a fundamentally different approach to problem-solving, which requires new mathematical frameworks and tools.

Another major challenge is the shortage of skilled professionals with the expertise needed to develop and implement quantum algorithms and applications. The development of quantum computing requires expertise in quantum mechanics, computer science, and mathematics. At present, there is a shortage of professionals with these skills, making it difficult to develop and implement quantum applications.

Conclusion: In conclusion, the potential impact of quantum computing on finance, healthcare, and transportation is enormous. Quantum computing has the potential to solve some of the most complex problems facing these industries, leading to more efficient and effective operations. However, there are significant challenges that need to be addressed before quantum computing can become a practical tool for solving real-world problems. With continued investment in research and development, we may soon see quantum computing become a key driver of innovation in these and other industries.

The ethical considerations of using quantum computing in AI

As we explore the exciting potential of quantum computing in AI and machine learning, it is important to also consider the ethical implications of this revolutionary technology. With the power to process vast amounts of data quickly and accurately, quantum computing could bring about new challenges and risks that must be addressed.

One significant concern is the potential for increased surveillance and monitoring of individuals. As quantum computing algorithms enable faster and more efficient data processing, there is a risk that sensitive information could be collected and misused, particularly in law enforcement and national security contexts. This raises important questions about the balance between public safety and individual privacy rights.

Another issue is data privacy. Quantum computing algorithms could potentially bypass privacy protections, allowing for the collection and analysis of personal data without individuals' knowledge or consent. This could have serious implications for privacy rights and could lead to discrimination or bias in decision-making processes.

Additionally, the use of quantum computing in AI and machine learning could reinforce systemic biases and discrimination if the data used to train algorithms is not representative of diverse populations. This could have serious implications for marginalized communities and could exacerbate existing inequalities.

There is also the risk of unintended consequences. Despite the best intentions, quantum algorithms could potentially create new problems or exacerbate existing ones, such as environmental problems or labor exploitation.

Finally, the automation and displacement of jobs is a potential concern, as is the distribution of benefits and opportunities created by quantum computing. Ensuring that the benefits of this technology are distributed fairly and equitably will be an important challenge.

To address these ethical considerations, it is crucial to take a proactive and collaborative approach. This may involve developing guidelines and standards for responsible use of quantum computing in AI and machine learning, as well as involving a diverse range of stakeholders in the development process. It is also important to invest in education and training programs to prepare individuals for the potential opportunities and challenges created by quantum computing.

Overall, the ethical considerations of quantum computing in AI and machine learning must be carefully considered as we explore the exciting potential of this revolutionary technology. By taking a proactive and responsible approach, we can ensure that the benefits of quantum computing are realized in a way that is fair and equitable for all.

Case studies of current research and applications of quantum computing

Quantum Chemistry Simulations

Quantum chemistry simulations are perhaps the most well-known and developed application of quantum computing in the field of AI and machine learning. These simulations involve modeling and simulating the behavior of molecules and chemical reactions, which is an incredibly complex task that classical computers are not well-equipped to handle.

Quantum computers, on the other hand, are ideally suited for these simulations due to their ability to model and simulate quantum mechanical behavior. By accurately simulating the behavior of electrons and atoms within a molecule, quantum computers can provide much more accurate and efficient calculations of molecular properties.

One example of a quantum chemistry simulation is the simulation of the nitrogenase enzyme, which is involved in the process of nitrogen fixation. Nitrogen fixation is the process by which atmospheric nitrogen is converted into ammonia, which is an essential component of many fertilizers.

The nitrogenase enzyme is a complex molecule that is difficult to study using traditional methods, but quantum chemistry simulations have provided new insights into its behavior and could help researchers develop more efficient and effective ways to promote nitrogen fixation.

Quantum Deep Learning

Another exciting area of research and application for quantum computing in AI and machine learning is in the development of quantum deep learning algorithms. Deep learning is a powerful approach to machine learning that involves training neural networks on large datasets to recognize patterns and make predictions.

However, traditional deep learning algorithms can be computationally expensive and require large amounts of processing power to train. Quantum deep learning algorithms offer the potential to significantly improve the efficiency and accuracy of these algorithms.

One example of a quantum deep learning algorithm is the quantum Boltzmann machine, which is a type of quantum neural network. This algorithm uses quantum mechanics to model the behavior of large groups of neurons, providing more efficient and accurate training of neural networks.

Another example is the quantum support vector machine, which is a type of quantum machine learning algorithm that is designed for classification and regression tasks. This algorithm uses quantum mechanics to model the behavior of data points in a high-dimensional space, providing more accurate and efficient classification and regression.

Ethical Considerations

As with any new technology, there are ethical considerations to be aware of when it comes to using quantum computing in AI and machine learning. One potential concern is the increased surveillance capabilities that quantum computing could provide. As quantum computers become more powerful, they could be used to break encryption and compromise sensitive data.

Another concern is the potential impact on employment. As quantum computing and AI become more advanced, there is a risk that many jobs could be automated or eliminated, which could have significant social and economic implications.

It is important for researchers and developers to consider these ethical concerns and work to ensure that quantum computing is used in a responsible and beneficial way. This could involve developing new encryption techniques that are resistant to quantum computing attacks or investing in education and training programs to help workers transition to new jobs in a changing economy.

In conclusion, quantum computing is an exciting and rapidly evolving field that has the potential to revolutionize AI and machine learning. From quantum chemistry simulations to quantum deep learning algorithms, there are a wide range of promising applications for this technology.

However, there are also challenges and ethical considerations to be aware of, including the need for error correction and decoherence management, as well as the potential for increased surveillance and the impact on employment.

As quantum computing continues to evolve and mature, it is important for researchers and developers to work together to ensure that this technology is used in a responsible and beneficial way. By doing so, we can unlock the full potential of quantum computing and create a more innovative and equitable future for all.

CONCLUDING REMARKS

Intelligence, at last?

The Turing test is a test proposed by the renowned British mathematician and computer scientist, Alan Turing, in 1950. It was designed to assess a machine's ability to exhibit intelligent behavior indistinguishable from that of a human. The test serves as a benchmark for evaluating the development of artificial intelligence (AI).

Alan Turing proposed the test in his paper titled "Computing Machinery and Intelligence." In the paper, he raised the question, "Can machines think?" and argued that the question is not about whether machines can replicate human thinking exactly but rather whether they can mimic human-like intelligent behavior convincingly enough to fool an observer.

The basic premise of the Turing test involves a human judge who engages in a natural language conversation with two entities—a human and a machine—without being aware of which is which. The judge's task is to determine which entity is human and which is the machine. If the machine can consistently deceive the judge into believing it is human, then it is said to have passed the Turing test.

Turing suggested that for the test to be fair, the machine should not be limited to responding to specific questions or tasks but should have the ability to engage in unrestricted conversation, demonstrating its understanding, knowledge, reasoning, and ability to generate human-like responses.

The conversation typically takes place through a text-based interface to remove any biases related to physical appearance or voice. The judge is allowed to ask questions or present challenges on any topic, and both the human and the machine respond in text form, without any visual or auditory cues.

Since its inception, the Turing test has sparked considerable debate among scientists, philosophers, and AI researchers. Many variations and criticisms of the test have emerged over the years. Some argue that passing the Turing test does not necessarily equate to true intelligence but rather the ability to imitate intelligence. Others believe that the test sets an important goal for AI development and serves as a practical benchmark to evaluate progress.

Various competitions and events have been organized to put AI systems to the test. One notable example is the Loebner Prize, an annual competition where chatbot programs compete to convince human judges that they are human. However, no machine has yet passed the Turing test with absolute success, though some chatbots have come close by convincing a significant portion of judges.

The Turing test has played a crucial role in the development of AI. It has encouraged researchers to focus on natural language processing, machine learning, and other areas essential for creating conversational AI systems. Although the test remains a subject of ongoing discussion and criticism, it continues to be a significant milestone in the field of artificial intelligence, driving progress toward achieving more human-like machine intelligence.

However, I believe that the following stories would convince you that even though the Turing test is absolutely fundamental for AI and its influence on humanity, it does not capture the entire picture.

A fish or no fish

A team of researchers embarked on training an AI system to recognize different species of fish. Their goal was to develop a robust system that could accurately identify fish species by analyzing images of them. To achieve this, they gathered a vast dataset consisting of thousands of images featuring various species of fish captured in different settings and contexts.

The researchers carefully labeled each image, indicating the specific fish species present in the picture. They used this labeled dataset to train the AI system, employing advanced computer vision techniques and machine learning algorithms. The aim was to teach the AI to identify the distinguishing features and patterns associated with different fish species.

The training process began, with the AI system being exposed to countless images of fish. The researchers expected the system to gradually learn the unique characteristics of each fish species, such as their color, scale patterns, and body shape.

However, as the researchers evaluated the AI system's progress, they stumbled upon a surprising discovery. They found that the AI was consistently identifying fish species, not based on their inherent traits, but by recognizing the presence of a human hand in the images.

Upon closer inspection, they realized that a significant portion of the training dataset contained pictures of people holding fish. The AI had inadvertently learned to associate the presence of a human hand with the likelihood of a fish being present. It had identified a strong correlation between fish recognition and the context of a human holding them.

This unexpected outcome led to a humorous situation where the AI system was successfully detecting fish species, but primarily by recognizing the human hand in the image. Essentially, the AI had learned to associate the fish with the accompanying hand rather than focusing on the fish itself.

Looking up

In 2015, a team of researchers at Google embarked on a project to improve the accuracy of their image recognition algorithms, specifically in identifying sheep. They aimed to train a neural network using a vast dataset of images that included various pictures of sheep in different environments.

The researchers painstakingly labeled each image, indicating whether or not it contained a sheep. They expected the neural network to learn and recognize the visual patterns and features associated with sheep, such as their wool, shape, and distinct facial characteristics.

The training process commenced, with the neural network being exposed to countless images of sheep. The researchers eagerly awaited the network's progress, anticipating that it would become proficient in identifying sheep accurately.

However, as they evaluated the performance of the trained neural network, they made a surprising discovery. The network was indeed capable of recognizing sheep, but it had also formed an unexpected association between clouds and sheep.

Further investigation revealed that many of the images in the training dataset featuring sheep had a common characteristic: they were taken against a backdrop of cloudy skies. The neural network had inadvertently learned to associate the presence of clouds with the presence of sheep, assuming a correlation between the two.

As a result, when the researchers tested the neural network with new images, it often classified images containing clouds as images of sheep. The network had learned to identify not only the sheep but also the clouds as a defining feature of sheep, despite the researchers' original intent to solely focus on sheep recognition.

This peculiar outcome provided a lighthearted and unexpected twist to the project. While the researchers had aimed to enhance the recognition of sheep, they unintentionally ended up with a neural network that associated clouds with sheep due to the prevalence of cloudy backgrounds in the training images.

While amusing, these incidents also highlighted an important lesson in AI comparison with human intelligence: would you make these mistakes, at any point throughout your life? If AI is truly a man-made intelligence, driven from the way humans learn, then why does it make mistakes we would never make? Are these creatures actually intelligent like us, or are we simply designing them to fool us behind closed doors?